A Gentle Introduction to Scientific Computing

Numerical Analysis and Scientific Computing Series

Series Editors:

Frederic Magoules, Choi-Hong Lai

About the Series
This series, comprising of a diverse collection of textbooks, references, and handbooks, brings together a wide range of topics across numerical analysis and scientific computing. The books contained in this series will appeal to an academic audience, both in mathematics and computer science, and naturally find applications in engineering and the physical sciences.

Handbook of Sinc Numerical Methods
Frank Stenger

Computational Methods for Numerical Analysis with R
James P Howard, II

Numerical Techniques for Direct and Large-Eddy Simulations
Xi Jiang, Choi-Hong Lai

Decomposition Methods for Differential Equations
Theory and Applications
Juergen Geiser

Mathematical Objects in C++
Computational Tools in A Unified Object-Oriented Approach
Yair Shapira

Computational Fluid Dynamics
Frederic Magoules

Mathematics at the Meridian
The History of Mathematics at Greenwich
Raymond Gerard Flood, Tony Mann, Mary Croarken

Modelling with Ordinary Differential Equations: A Comprehensive Approach
Alfio Borzì

Numerical Methods for Unsteady Compressible Flow Problems
Philipp Birken

A Gentle Introduction to Scientific Computing
Dan Stanescu, Long Lee

For more information about this series please visit: https://www.crcpress.com/Chapman-HallCRC-Numerical-Analysis-and-Scientific-Computing-Series/book-series/CHNUANSCCOM

A Gentle Introduction to Scientific Computing

Dan Stanescu
University of Wyoming, USA

Long Lee
University of Wyoming, USA

CRC Press
Taylor & Francis Group
Boca Raton London New York

CRC Press is an imprint of the
Taylor & Francis Group, an **informa** business

A CHAPMAN & HALL BOOK

First edition published 2022
by CRC Press
6000 Broken Sound Parkway NW, Suite 300, Boca Raton, FL 33487-2742

and by CRC Press
4 Park Square, Milton Park, Abingdon, Oxon, OX14 4RN

© 2022 Taylor & Francis Group, LLC

CRC Press is an imprint of Taylor & Francis Group, LLC

Library of Congress Cataloging-in-Publication Data

Names: Stanescu, Dan, 1982- author.
Title: A gentle introduction to scientific computing / Dan Stanescu,
   University of Wyoming, USA, Long Lee, University of Wyoming, USA.
Description: First edition. | Boca Raton : Chapman & Hall, CRC Press, 2022.
   | Series: Chapman & Hall/CRC numerical analysis and scientific computing
   series | Includes bibliographical references and index.
Identifiers: LCCN 2021056202 (print) | LCCN 2021056203 (ebook) | ISBN
   9780367206840 (hardback) | ISBN 9781032261317 (paperback) | ISBN
   9780429262876 (ebook)
Subjects: LCSH: Numerical analysis--Data processing. | Computer science. |
   Science--Data processing.
Classification: LCC QA297 .S695 2022 (print) | LCC QA297 (ebook) | DDC
   518.0285--dc23/eng20220301
LC record available at https://lccn.loc.gov/2021056202
LC ebook record available at https://lccn.loc.gov/2021056203

ISBN: 9780367206840 (hbk)
ISBN: 9781032261317 (pbk)
ISBN: 9780429262876 (ebk)

DOI: 10.1201/9780429262876

Typeset in Latin Modern font
by KnowledgeWorks Global Ltd.

Publisher's note: This book has been prepared from camera-ready copy provided by the authors.

Contents

Preface

There exists a large variety of books dealing with computational methods and the material included here is therefore by no means new or original. The natural question that arises is then: why a new book? The simple answer is that the presentation here is driven by personal experience; the material is looked at in a way that has been found the most useful when teaching it to a large body of students over more than fifteen years. Together, the authors' work in the field spans around five decades, with more than thirty years of experience in teaching numerical methods. In terms of mathematical complexity, the approach is somewhat of a middle ground between the more involved and very systematic, while more heavily theoretic, presentation found in texts like M.H. Holmes [7] or M.T. Heath [6] for example, and the very applied but less mathematically-oriented point of view taken by D.T. Kaplan [10]. One of the other points where this presentation differs from others might also be the choice to place an emphasis on computational efficiency. Many computational scientists, mathematicians in particular, eventually devote themselves to the more academic task of developing proof-of-concept programs that use low-dimensional toy models to show how the computation may proceed. While this is a very important endeavor, the truth is that eventually we need computer models that track real-world phenomena, like for weather prediction, component design and optimization or turbulent flow simulation. For such real applications, which can easily still keep even the largest computers available today busy for days, weeks or even years in a row, computational efficiency is crucial. For this reason, this material emphasizes, to the largest extent possible, some simple techniques that can make a large difference in computer time. They are definitely worth being learned from the very beginning so that thinking about efficiency becomes second nature.

A class at the junior level has been taught at the University of Wyoming based on some initial notes, out of which this book has slowly grown out over the years. Other, sometimes more advanced, courses in numerical methods were a staple in the mathematics department as well as other departments across the University of Wyoming campus. They all took for granted a previous exposure to a computer science course, such as Introduction to Coding. By contrast, when initially designed, this junior-level class had as a particular objective to introduce mathematics majors, without any prior computer science exposure whatsoever, to numerical methods. Initially taught to sections of about ten students every other semester, the class now runs every semester, with two full sections totaling around seventy students in the Fall semester and one in Spring. About half of the students are mathematics (pure or education) majors, with the large majority of the rest coming from engineering. The students are expected to have a knowledge of multivariable calculus; no initial knowledge of linear algebra or differential equations is required. To compensate, one of the initial chapters covers the most important linear algebra concepts. These are dealt with completely during the first week of classes, although some reinforcement is sought throughout the semester via homework assignments. On the other hand, the basics of ordinary differential equations are presented before pertinent numerical methods are introduced in Chapter 10, which is dedicated to this topic. This latter chapter is usually the last one addressed; the better part of two weeks is spent on the topic, with at least one lecture focusing on solving boundary value problems. This choice is motivated by the fact that the shooting method is a good final project choice: it brings together a couple of topics visited throughout the semester (i.e. solving nonlinear equations and differential equations) while also inviting students to exercise their capacity for generalization and abstraction.

For a book dealing with computer programming, conventions are mandatory; they help separate computer commands from the rest of the text. In this book, every effort has been made to identify important terms and concepts that also appear in the index. When such terms are first introduced, they appear in italics. MATLAB® commands appearing throughout the text will be in true type font, like this: `plus`. Snippets of computer code that can be used to perform a stand-alone task (for example, solving an equation or computing an interpolating polynomial) will be placed inside special boxes. These will be, most of the time, either MATLAB scripts or MATLAB functions and start appearing in Chapter 4. The text doesn't always provide complete code for all the methods discussed herein. Instead, code is provided for many basic parts, while readers are implicitly encouraged to assemble these parts into final code themselves. Many times, this process is supported by previous exercises that appeared in previous chapters. As an example, a code for computing the spline coefficients is included, but students are left to write code that evaluates the spline interpolant. This is just a piecewise function; simpler versions of piecewise functions already appear in the exercises that follow the chapter on MATLAB basics. Careful planning will help instructors build students' knowledge of MATLAB, together with that of numerical methods, in an incremental, as-effortless-as-possible way.

Chapter 1

Introduction

1.1 Scientific Computing

This book aims to expose you to some of the basic tools and techniques used in Computational Science (CS henceforth). A relatively new, interdisciplinary area of science having its roots primarily in Applied Mathematics, Computer Science, Physics and Statistics, but encompassing a much larger spectrum of disciplines, from life sciences to engineering and the world of finance, CS may be broadly defined as the accumulated body of knowledge related to the particular use of computers that aims at gaining scientific understanding and insight. The former President B. Obama's Information Technology Advisory Committee (PITAC) proposed a rather broad definition of CS. According to the PITAC report [3], CS is a rapidly growing multidisciplinary field that uses advanced computing capabilities to understand and solve complex problems by fusing three distinct elements: (a) algorithms and modeling and simulation software, (b) computer and information science and (c) the computing infrastructure. The report further states that CS now constitutes what many call the third pillar of the scientific enterprise, a peer alongside theory and physical experimentation. Because most scientific computational tasks, regardless of the background discipline, eventually boil down to several similar steps (from setting up the encompassing algorithm to opening and inspecting data files, solving equations and other mathematical manipulations to finally visualizing and interpreting the results) a well-trained computational scientist may be expected to eventually be able to use his skills just as easily to compute forces on the structure of a building as to model the path of a cyclone or the expected price of options on the stock market. Since our purpose is to understand what happens behind the scenes when simple tools such as solving equations, plotting and interpolation are used, we will explore them both from a mathematical perspective (taking a peek at their mathematical foundations) and from an implementation, programming perspective. The latter is very important: having been exposed to mathematical concepts and notation for a long time, many of us in the exact sciences have more or less incorporated them into our thought processes. However, that doesn't mean that everyone is automatically able to convey them in a way that can be used by a computer. One may fail to express these concepts algorithmically, as a series

DOI: 10.1201/9780429262876-1

of steps that need to be taken in succession to solve a given problem, which is simply the way a computer works. While the terms Computational Science and Scientific Computing are sometimes used interchangeably, we will restrict our definition of Scientific Computing to encompass that area of Computational Science that focuses primarily on basic, mathematics-rooted numerical techniques and methods that are usually needed for completing computational tasks, essentially a part of the first element in the PITAC definition.

1.2   MATLAB®: What and Why?

MATLAB® is a programming environment that has emerged as a tool of choice for applied mathematicians and scientists involved with computation over the last decade. The name MATLAB is, to the best of author's knowledge, an abbreviation for "matrix laboratory"; initially MATLAB was conceived as a tool to handle matrices and vectors with ease. MATLAB is not a programming language proper; instead it is a programming environment that contains several components. The most important ones are: the interpreter (which takes the user's programming commands in the command window and outputs the results immediately), the editor (which allows you to type nicely-formatted code and save it as script and function files), the graphics package (which takes computed data and displays them for visual inspection), and the file system handler (which opens and closes files and manages all pertaining interactions with the operating system). As opposed to a programming language, MATLAB does not compile your programs but interprets your commands one by one and outputs the results as soon as they are available. This simplifies the task of finding errors in a piece of code (still, as you'll see, this task is many times not a trivial one) while ultimately making a sacrifice on efficiency. Because MATLAB is a good compromise between a high level programming language and the mathematical notation that we are used to, and because of its widespread use, it has been chosen as the proper tool for conveying the concepts in this text.

While largely used, MATLAB is not free; a license must be purchased. If the price of the software is a major drawback, a free alternative is available: OCTAVE is an open-source program that can be downloaded from the Internet, installed and used for free on anyone's personal computer. It does offer a couple of options: a graphical user interface (GUI) similar to MATLAB and a command-line interpreter (CLI) version that is much faster albeit less convenient. For those willing to explore a little further, Python is another high-end language with free implementations that can be either installed or (a variant many prefer) run in a browser window. Python has raised in the last decade as the programming language of choice for data analysis, data science and machine learning (while these are mentioned like three different fields, there is a strong overlap in between them) and has also almost completely replaced other languages (i.e. Java) as the first language taught to Computer Science majors in US colleges. Its appeal comes not only from its high readability and flexibility, by themselves two somewhat antagonistic aspects, but also from the large array of packages and libraries that can be used on-the-fly with very little or no experience. Due to the popularity of the language, there is also a large number of books that teach Python, some of them available as free resources over the Internet. A very good example that actually deals with the same topic as this book while teaching Python from scratch is the text by J. Kiusalaas [11]. An appendix replicating some of the MATLAB codes developed throughout the following pages into Python codes is also provided at the end. A big drawback of MATLAB, however, as well as Python to some extent, is the execution speed. Because it is an interpreted language, its performance is highly reduced, for example because of continuously waiting for input from the user in an interactive session. In the realm of scientific computing this may turn out to be a big drawback, since many of our computational tasks (for example weather forecasting or modeling of turbulent fluid flow) still push the available computational resources to their limit. Therefore, MATLAB is widely used for proof-of-concept implementations and teaching, while high-end computational tools are mostly written in compiled languages. The computational science community, aware of both the advantages and disadvantages of such interpreters for a long time, recently started developing a different approach which takes somewhat of a middle path. One of the outcomes of such efforts is the Julia programming language. Julia uses a notation very similar to MATLAB and has largely the same high level of mathematical capability, so that someone familiar with the latter language can easily make the transfer. Nevertheless, because of pre-compilation techniques used in Julia, some benchmarks show that it outperforms MATLAB in terms of computing time by as much as a factor of ten. Just like Python, Julia is also available either as a free download and local install or in a web browser. A more in-depth presentation, together with some example Julia codes are again provided in a separate appendix.

1.3   A Word of Caution

Numerical analysis as a branch of mathematics is well established. A plethora of very well-written, in-depth treatments of the field are available. Nonetheless, while not the rule, many of these don't offer all details on how to best implement the methods they deal with. One of the main goals of this book is to eventually enable its users to write useful computer code. This is not feasible without at least a basic understanding of the most basic numerical methods which are presented here. This is not enough, however; a different way of thinking about the problems to be solved, *algorithmic thinking*, already mentioned above, is required in order to interact with a computer. This is related to the capacity of expressing the task of reaching a goal (in this case obtaining the result of a computation) as a series of smaller steps which, when performed in succession, allow one to attain such goal. Like with all our skills, for some people this is more intuitive than it is for others. Nonetheless, we all can augment this skill by dedicated study and practice. However, dedicated programming practice is many times very difficult as it may turn into outraging frustration. When this happens, and it is almost guaranteed to happen, remember to take a break and possibly look for assistance. If no experienced friend or teacher is available, reach out on the Internet, which offers a lot of ways to get help. One can ask almost any question on Stack Exchange, for example. This is a huge advantage compared to the days when we used to walk into the computer building with a box of punched cards after waiting in line a couple of days to encode our program on them, then keeping our breath in expectation of the moment when our code will get to be run, usually within a week or so. A simple missed key, and the cycle would need to be repeated. Don't let frustration stop you from learning a very valuable skill, and remember that most of the time this experience is needed for us all to learn; it will eventually change into tremendous satisfaction once you have it all figured out.

1.4   Additional Resources

While the bibliography has been kept at a minimum, there are many freebies available on the Internet for learning MATLAB (or Python or Julia, if you so prefer). A simple Google search for "MATLAB tutorial" will reveal a vast array of documents, pieces of code and videos, from those produced by the MathWorks company itself (MATLAB's owners and developers) to many others written for scientific computing classes similar to this one. Since everyone learns in their own way and active involvement is encouraged, the readers are left to find the resources that fit them best. Nevertheless, two exceptional resources are worth mentioning for consideration. The first is the freely-available text *Numerical Computing with MATLAB* written by the original MATLAB developer Cleve Moler [15]. The second is Wikipedia; pages on numerical methods on this overarching website tend to have extensive information, sometimes including full-blown examples of MATLAB codes.

Chapter 2

Vectors and Matrices

Computers are particularly useful for performing repetitive tasks. In order to repeat the same kind of operation over and over again, but on different quantities, it is easiest to organize these quantities into indexed collections. The index, usually an integer, is used in order to gain access to the quantity it points to within a collection. Thus, each member or subset of a collection will be available for computing with, as soon as its index, or index set, is known. Vectors and matrices are two natural indexed collection types that are probably used more often than any other data types in scientific computer programming. Furthermore, they benefit from a well-established and rich mathematical background. While vectors and matrices are useful and may be encountered in a variety of contexts throughout several disciplines, this chapter will explore them from a rather linear algebra perspective. For a more in-depth presentation of these topics the reader is invited to consult one of the many available texts on linear algebra, for example S. Boyd [2]. The next chapter will focus on the creation and use of these collections within MATLAB.

2.1 Unidimensional Arrays: Vectors

A *vector* is defined here simply as an ordered collection of elements, alternately known as its entries. For the purpose of this book, it will be sufficient to think about these elements simply as holding numerical values. Vectors will be denoted henceforth by boldface characters and are usually represented with their entries listed in between brackets (either square or curved) in a vertical column. For example,

$$\mathbf{x} = \begin{bmatrix} 1.1 \\ 2.3 \\ -0.5 \end{bmatrix}$$

is a vector with three elements. The individual entries in the vector are available through their index, which in usual mathematical notation ranges from one to the number of elements in a vector. Thus, we can write $x_1 = 1.1$ and $x_3 = -0.5$ for the first and last elements of the vector $\mathbf{x}$. Note that the entries are here denoted by usual typeface, as opposed to the bold font used for the

DOI: 10.1201/9780429262876-2

5

vector as a whole object; the index appears as a subscript, hence x_i is the i-th element in $\mathbf{x}$. If the number of entries in $\mathbf{x}$ is n, the vector is known as an n-vector. It is common to refer to the entries in a vector as scalars. For most of our purposes these scalars are just real numbers, although sometimes it may be more convenient to restrict them to integers or, on the contrary, allow them to be complex numbers.

One can think about the same 3-vector, now with a new name $\mathbf{y}$ for convenience, as a horizontal row of entries, possibly separated by commas to avoid confusion:

$$\mathbf{y} = (1.1, 2.3, -0.5).$$

However, in order to effect an easier transition between this more flexible mathematical notation and the somewhat more rigid norms imposed by MAT-LAB, it is useful to differentiate between column vectors (aligned vertically) and row vectors (aligned horizontally). Thus, while it is true that $x_i = y_i$ for all values of i in the proper range of the index, i.e. for $i = 1, 2, 3$, so the two vectors are equal element-wise, they will be considered different kinds of objects. As will be seen in the next section, they are both different instances of a more general two-dimensional array: a matrix. As such, some operations will not be allowed or even defined in between the two different vector types (for example, addition) while others are (multiplication, which in this case becomes the dot product). Nevertheless, a row vector can be converted into a column vector by an operation known as transposition and denoted by a right-superscript letter T; thus $\mathbf{x}^T = \mathbf{y}$ and $\left(\mathbf{x}^T\right)^T = \mathbf{y}^T = \mathbf{x}$. As a final remark, please be aware that many linear algebra texts will consider vectors to always be column vectors; in that case the row vectors need to consistently be denoted with the transcript symbol. Although this is the mainstream convention and as far as possible we'll abide to it, it is somewhat inconvenient when interacting with MATLAB, for which the default vector type is a row vector.

2.2 Bidimensional Arrays: Matrices

A *matrix* is a rectangular array of scalars, arranged in rows and columns. Matrices will be denoted by capital letters, while their entries will be denoted by the corresponding lower-case letter. For example, the matrix A with m rows and n columns has the entry a_{ij} in the i-th row and the j-th column and may be conveniently represented in terms of its entries as $A = (a_{ij})$. In other words, the array of matrix entries has the form:

$$A = \begin{bmatrix} a_{11} & a_{12} & \cdots & a_{1n} \\ a_{21} & a_{22} & \cdots & a_{2n} \\ & \vdots & & \\ a_{m1} & a_{m2} & \cdots & a_{mn} \end{bmatrix}$$

To emphasize the size of the matrix A, we say that it is an $m \times n$ matrix and denote it as $A(m \times n)$. If the number of rows m is equal to the number of columns n, the matrix is known as a square matrix; square matrices are most common in applications. A square matrix of size $(n \times n)$ is usually known as an order-n matrix.

Related to any matrix A are two other matrices, the *transpose* of A, denoted by A^T and defined as $A^T = (a_{ji})$ and the transpose of the complex conjugate, also known as the *adjoint* of A, denoted here by $A^\dagger = (\bar{a}_{ji})$. We used the overbar notation for the complex conjugate of a scalar with real part x and imaginary part y. That is, if $a_{km} = x + iy$, then $\bar{a}_{km} = x - iy$, where $i = \sqrt{-1}$. It follows that the adjoint is the same as the transpose i.e. $A^\dagger = A^T$, if and only if all the entries in A are real. Furthermore, notice that a row n-vector is just a $(1, n)$ matrix, while a column vector of the same size is a $(n, 1)$ matrix. A square matrix for which $A^T = A$ is known as a *symmetric matrix*, while a square matrix for which $A^\dagger = A$ is known as *Hermitian*. Obviously, a symmetric matrix with real-valued entries is also Hermitian.

Example Given the matrix

$$A = \begin{bmatrix} 3+i & 2+3i & 1-i \\ 5-2i & 5-i & 1+i \end{bmatrix},$$

its transpose and adjoint are:

$$A^T = \begin{bmatrix} 3+i & 5-2i \\ 2+3i & 5-i \\ 1-i & 1+i \end{bmatrix}, \quad A^\dagger = \begin{bmatrix} 3-i & 5+2i \\ 2-3i & 5+i \\ 1+i & 1-i \end{bmatrix}.$$

$\parallel$

2.3 Matrix Operations

Most matrix operations are defined as expected, element-wise; multiplication is however an important exception. To start with, two matrices are equal if and only if they have the same size and the corresponding matrix elements are equal. The important matrix operations are listed here:

Addition: The sum of two matrices A and B is only defined if they have the same size, say $m \times n$. Then the elements of the matrix $C = A + B$ are given by $c_{ij} = a_{ij} + b_{ij}$, for $i = 1, 2, \ldots, m$ and $j = 1, 2, \ldots, n$.

Multiplication by a scalar: The product of a matrix A with a (possibly complex) number z is the new matrix $zA = (za_{ij})$; i.e. all entries are multiplied by the scalar. If the scalar is $z = -1$ then the matrix $B = (-1)A = -A$ and $A + B = B + A = \mathbf{0}$. The symbol $\mathbf{0}$ thus denotes the null matrix for addition, i.e. $A + \mathbf{0} = \mathbf{0} + A = A$. The size of $\mathbf{0}$ is often determined from the context and will need to be specified otherwise. Subtraction of two different matrices of the same size is of course defined accordingly: $A - B = A + (-B)$.

Matrix Multiplication: The product of two matrices $A(m \times n)$ and $B(n \times p)$ is defined as $C = AB = (c_{ij})$ where

$$c_{ij} = \sum_{k=1}^{n} a_{ik} b_{kj}, \quad i = 1, 2, \ldots, m; \ j = 1, 2, \ldots, p.$$

As mentioned previously, this immediately defines the product of a row vector with a column vector of the same size. Thus, suppose that $\mathbf{x}$ and $\mathbf{y}$ are two column n-vectors; then their *dot product* is given by:

$$\mathbf{x}^T \mathbf{y} = \sum_{k=1}^{n} x_k y_k.$$

The dot product may be a complex number if at least one of the vectors has complex entries. A different product, the *inner product*, can be defined as an extension of the dot product:

$$(\mathbf{x}, \mathbf{y}) = \sum_{k=1}^{n} \bar{x}_k y_k.$$

Because the product of a complex number with its complex conjugate is always a non-negative real quantity, the inner product of a vector $\mathbf{x}$ with itself satisfies $(\mathbf{x}, \mathbf{x}) = \sum_{k=1}^{n} x_k \bar{x}_k \geq 0$. Thus, the inner product of a vector with itself will always be a non-negative real number, even if the entries in the vector $\mathbf{x}$ are complex. Moreover, the equality only holds if all the elements of the vector $\mathbf{x}$ are zero: $(\mathbf{x}, \mathbf{x}) = 0$ if and only if $\mathbf{x} = \mathbf{0}$. This non-negativity allows us to define a quantity known as the 2-norm of a vector $\mathbf{x}$ by:

$$\|\mathbf{x}\|_2 = \sqrt{(\mathbf{x}, \mathbf{x})}.$$

Obviously, if the two vectors $\mathbf{x}$ and $\mathbf{y}$ have only real-valued entries then their dot product coincides with the inner product. If two vectors satisfy $(\mathbf{x}, \mathbf{y}) = 0$, they are said to be *orthogonal vectors*. Notice also that for a 3-vector, the 2-norm

$$\|\mathbf{x}\|_2 = \sqrt{(\mathbf{x}, \mathbf{x})} = \sqrt{x_1^2 + x_2^2 + x_3^2}$$

is the Euclidean distance from the tip of the vector to the origin, also known as the length of the vector.

Matrix multiplication is not commutative: in general, $AB \neq BA$. In fact, if A is a matrix of size $(m \times n)$ and B a matrix of size $(n \times p)$ with $m \neq p$, then the product AB is defined, but BA is not. It can be verified by direct, albeit tedious, calculation that matrix multiplication is distributive with respect to matrix addition, i.e. $A(B + C) = AB + AC$ and associative i.e. $A(BC) = (AB)C$. A special matrix denoted by I which has all the entries equal to zero except for those on the main diagonal,

$$ I = \begin{bmatrix} 1 & 0 & \cdots & 0 \\ 0 & 1 & \cdots & 0 \\ \vdots & & & \\ 0 & 0 & \cdots & 1 \end{bmatrix}. $$

is the *identity matrix* for multiplication. That is, for any square matrix A multiplication by this matrix is commutative and leaves the matrix A unchanged: $AI = IA = A$.

Once the identity matrix is available, one can in principle define a multiplicative inverse for a square matrix A of order n, denoted by A^{-1} and satisfying $AA^{-1} = A^{-1}A = I$. However, such inverse does not exist for all square matrices; the matrices for which it does not exist are called *singular*. Nonsingular matrices, on the other hand, admit an inverse. It can be shown that the inverse of a nonsingular matrix is unique. Singularity of a matrix is closely related to the value of a scalar quantity that can be computed for each matrix A from its entries a_{ij}. This quantity is known as the *determinant* of that matrix and is denoted by $\det(A)$.

2.4 Systems of Linear Equations

Suppose one needs to find the point of intersection of two straight lines in a plane where a generic point P is located by its coordinates (x, y). The equation of a straight line in the plane has the form $y = \alpha x + \beta$ or alternately $a_1 x + a_2 y = b$. The point of intersection, assuming the two lines are not parallel, must satisfy the equations of both the first and the second line, simultaneously. Therefore, the coordinates of the intersection point must satisfy what is known as a two-by-two system of linear algebraic equations:

$$ \begin{cases} a_{11}x + a_{12}y &= b_1 \\ a_{21}x + a_{22}y &= b_2 \end{cases} \tag{2.1} $$

The first subscript index on the coefficients is used here to indicate the equation number, while the second is used to indicate the variable, either x or y.

Note that this can be written in matrix form as $A\mathbf{x} = \mathbf{b}$, where the matrix A is known as the matrix of the system,

$$A = \begin{bmatrix} a_{11} & a_{12} \\ a_{21} & a_{22} \end{bmatrix}.$$

The vector of unknowns is $\mathbf{x} = (x, y)^T$, the right-hand side vector is $\mathbf{b} = (b_1, b_2)^T$ and the equality is not assumed to be obeyed (i.e. it does not hold) for any $\mathbf{x}$, but rather it is asked for.

There are many ways one can think of for solving such a system. We'll explore here one method that can be systematically extended to systems of larger sizes: Gauss elimination. Let us start by noticing that multiplying either equation by a non-zero scalar will alter neither that particular equation nor the system. Let us suppose that a_{21} is not zero; otherwise one could readily solve for y from the second equation and subsequently use this value in the first equation to solve for x. One can proceed in the same manner if $a_{11} = 0$ upon switching the order of the two equations, if necessary. Let's therefore assume that $a_{11} \neq 0$ and let $m_1 = a_{21}/a_{11} \neq 0$. One can use this multiplier factor to change the first equation:

$$\begin{cases} a_{21}x + m_1 a_{12}y &=& m_1 b_1 \\ a_{21}x + a_{22}y &=& b_2 \end{cases}$$

Next, subtract this new first equation from the second to obtain the equivalent system:

$$\begin{cases} a_{21}x + m_1 a_{12}y &=& m_1 b_1 \\ 0x + (a_{22} - m_1 a_{12})y &=& (b_2 - m_1 b_1) \end{cases}$$

whence we can easily get the value of the y coordinate of the intersection point by solving the second equation thus obtained. We will denote this value by y^*, to indicate that this is the solution and not a generic variable:

$$y = y^* = \frac{b_2 - m_1 b_1}{a_{22} - m_1 a_{12}} = \frac{a_{11} b_2 - a_{21} b_1}{a_{11} a_{22} - a_{12} a_{21}},$$

a value defined as long as the denominator of the fraction is not zero. Using this value in the first equation then produces a similar expression for the x coordinate of the intersection point,

$$x = x^* = \frac{a_{12} b_2 - a_{22} b_1}{a_{11} a_{22} - a_{12} a_{21}}.$$

The denominator is exactly the *determinant* of the system matrix, usually denoted by listing the matrix entries in between vertical bars – for most of us a form that is easier to remember:

$$\det(A) = \begin{vmatrix} a_{11} & a_{12} \\ a_{21} & a_{22} \end{vmatrix} = a_{11} a_{22} - a_{21} a_{12}.$$

Notice that the determinant is a scalar (the seemingly array notation in between vertical bars should not mislead you). Also note that the value of the determinant for this (2×2) matrix is obtained by subtracting from the product of the two scalars on the main diagonal (running upper left to lower right) the product of the two terms on the other diagonal (upper right to lower left). The solution procedure will be successful if the matrix is nonsingular.

Let us now see how this elimination procedure works for a (3×3) system which might be interpreted as finding the point of intersection of three planes. Of course there will be situations when the three planes will not intersect in a point, but we will leave those aside for now. To simplify keeping track of the numbers obtained through the elimination procedure, we'll use a less generic example, with the matrix entries set to particular numbers. We also need a more systematic way to denote the multipliers, since there will be several such multipliers involved. Thus, consider the following system $A\mathbf{x} = \mathbf{b}$:

$$\begin{cases} x + y - z & = & 3 \\ 2x + 3y + z & = & 4 \\ 3x + 5y + 7z & = & 1 \end{cases}$$

The first phase in the Gauss elimination procedure is called *forward elimination*. In this stage one uses non-zero entries called pivots to obtain zero coefficients, with the aim of reducing the system to smaller and smaller sizes. The first pivot in this case is the x-coefficient in the first equation; it will be used to produce zeros in the first column of the system matrix, below the pivot. Multiplying the first equation with the two multipliers $m_{21} = 2$ and $m_{31} = 3$ and subtracting the two equations obtained by multiplying respectively with these two values from the second and third equations (while leaving the first equation unchanged), one obtains the new system:

$$\begin{cases} x + y - z & = & 3 \\ 0x + y + 3z & = & -2 \\ 0x + 2y + 10z & = & -8 \end{cases}$$

Notice that the multipliers are denoted with two indices, indicating the row and column where they are intended to produce a zero; thus m_{21} produces a zero in the second line, first column. Also notice that the last two equations now form a two-by-two system for the two variables y and z, which (except for the variable names) is similar to the one in equation 2.1. Proceeding in the same manner, one now uses the coefficient of y in the second equation (the new pivot) to obtain an additional zero in the last equation; the multiplier needed in this case is $m_{32} = 2$. Subtracting again the second line from the third upon multiplication produces:

$$\begin{cases} x + y - z & = & 3 \\ 0x + y + 3z & = & -2 \\ 0x + 0y + 4z & = & -4 \end{cases}$$

At this point the forward elimination stage is finished. The next stage is *backward substitution* and involves solving for the variables, proceeding from the last (in this case z) to the first (our x). The last equation produces $z = z^* = -1$; using that in the second equation one gets $y = y^* = 1$ and finally, by using these two values used in the first equation one gets $x = x^* = 1$. So the solution of the system is $\mathbf{x} = \mathbf{x}^* = (1, 1, -1)^T$. Again, we have used a star subscript to clearly denote the solution, as opposed to the generic variable $\mathbf{x}$; this is not a very common practice as most texts dealing with the problem don't make this distinction in an explicit way. However, although it can usually be understood from the context whether one refers to a generic point $\mathbf{x}$ or to the sought-after solution of the system, our notation for the solution will be helpful in the context of iterative methods.

Let us return to the solution process. The original system $A\mathbf{x} = \mathbf{b}$ was changed through the forward elimination step into the related system $U\mathbf{x} = \mathbf{v}$ where the matrix U is an upper-triangular matrix (i.e. has non-zero entries only above the main diagonal):

$$U = \begin{pmatrix} 1 & 1 & -1 \\ 0 & 1 & 3 \\ 0 & 0 & 4 \end{pmatrix}.$$

Upon some reflection one may realize that the matrix U is related to the matrix A by the relationship $A = LU$ where the matrix L is a lower-triangular matrix that contains the multipliers used in the elimination stage while its main diagonal has the entries equal to one:

$$L = \begin{pmatrix} 1 & 0 & 0 \\ m_{21} & 1 & 0 \\ m_{31} & m_{32} & 1 \end{pmatrix} = \begin{pmatrix} 1 & 0 & 0 \\ 2 & 1 & 0 \\ 3 & 2 & 1 \end{pmatrix}.$$

Writing the original system as $LU\mathbf{x} = \mathbf{b}$ and multiplying the two column vectors on the left- and right-hand sides of the equal sign with the inverse of the matrix L (which can be shown to exist in this case) one obtains $L^{-1}LU\mathbf{x} = L^{-1}\mathbf{b}$, the equality being kept as long as L^{-1} is not the zero matrix. Since $L^{-1}L = I$, this last equality is actually a restatement of the system obtained after forward elimination $U\mathbf{x} = \mathbf{v}$, where $\mathbf{v} = L^{-1}\mathbf{b}$. Looking at it this way, it may appear that one needs to compute the inverse L^{-1} in order to find the new right-hand side $\mathbf{v}$, but that is not the case because $\mathbf{v} = L^{-1}\mathbf{b}$ implies $L\mathbf{v} = \mathbf{b}$. Hence, as long as one knows the matrices L and U, one can solve this system $L\mathbf{v} = \mathbf{b}$ (with the system matrix L) to find $\mathbf{v}$, then solve the system $U\mathbf{x} = \mathbf{v}$ (with the system matrix U) to find the original unknown $\mathbf{x}$. Both these systems are considerably easier to solve than the original system. In fact, one can show that the number of arithmetic operations necessary for Gauss elimination grows as the third power is n if A is an order-n matrix, but only with the second power of n for the two associated triangular systems. On the other hand, the process of finding the two matrices L and U, known as the LU-factorization of the matrix A, is obviously equivalent to Gauss elimination in terms of its computational complexity.

Using the definition of the inverse, the solution of the system $Ax = b$ can be conveniently written in the form $\mathbf{x}^* = A^{-1}\mathbf{b}$, as long as the inverse of A exists. While this formula conceptually offers another way to compute the solution by first finding the inverse, this is not widely used in practical calculations because the computation of the inverse is itself equivalent to solving n systems of equations of the same size. Computation of the inverse should therefore be avoided whenever possible.

2.5   Eigenvalues and Eigenvectors

In the general case, multiplication of a matrix by a vector results in a new vector which has a different norm and also a different orientation. This can easily be seen for two-dimensional vectors, which can be represented in a plane. Consider for example the matrix:

$$A = \begin{bmatrix} 2.2 & -0.4 \\ -2.4 & 1.8 \end{bmatrix}$$

and the vector $\mathbf{v} = [2 \ 2]^T$ which is oriented along the first bisectrix; then the vector $A\mathbf{v} = [3.6 \ -1.2]^T$ not only has a different length, but is also oriented along a different direction than the original $\mathbf{v}$. However, for a given matrix A of order n there may be vectors for which multiplication with that particular matrix only leads to a change in norm, while the orientation remains the same. For example, multiplication of the vector $\mathbf{x} = [-1 \ 2]^T$ with the same matrix produces $A\mathbf{x} = [-3 \ 6]$ which is now a vector three times longer than $\mathbf{x}$ that is, however, oriented along the same line as $\mathbf{x}$. This can be seen in figure 2.1.

The vectors that keep their direction upon multiplication with a given matrix are known as the eigenvectors of that matrix, and play a very important role in understanding matrix multiplication. Let us denote such a vector by ξ; the condition that upon multiplication the direction of the vector is not changed can be written as

$$A\xi = \lambda\xi = \lambda I\xi$$

for some scalar λ, because multiplication of a vector by a scalar does not change the direction of the vector. Use has also been made of the fact that multiplication by the identity matrix I does not change the vector. Moving all the terms on the left-hand side, this equation can also be written:

$$(A - \lambda I)\xi = \mathbf{0},$$

which in turn gives us a way to find the vectors ξ. Notice first that the zero vector will always satisfy this equation for any value λ. The zero vector, however, does not specify a direction and is therefore useless in understanding

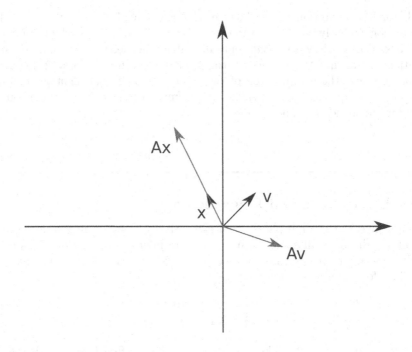

Figure 2.1: Action of a matrix on a general vector **v** compared to its action on an eigenvector **x**.

the action of the matrix A, so it makes sense to only be interested in the non-zero vectors that solve this equation. Because, as we have seen, a non-singular matrix has a definite inverse, if the matrix $A - \lambda I$ is non-singular, the only possible solution is the zero vector, also known as the *trivial solution*. For non-trivial solutions this matrix needs therefore to be singular. This condition can conveniently be written as:

$$\det(A - \lambda I) = 0,$$

an equation that is known as the *characteristic equation* of the matrix A. This is an algebraic equation of order n for the values λ, which are known as the eigenvalues of the matrix A. Such an equation will have n roots, although repeated roots are possible and some roots may be complex. For the matrix in our example the characteristic equation takes the form:

$$\det(A) = \begin{vmatrix} 2.2 - \lambda & -0.4 \\ -2.4 & 1.8 - \lambda \end{vmatrix}$$

$$= (2.2 - \lambda)(1.8 - \lambda) - 2.4 * 0.4 = \lambda^2 - 4\lambda + 3 = 0$$

which leads to the two roots that are the eigenvalues of A, $\lambda_1 = 1$ and $\lambda_2 = 3$.

In this particular case, both values are real. With these values one can proceed to find the eigenvectors. An important result from linear algebra [2] guarantees that, since the eigenvalues are distinct, there will be two different eigenvectors in this case. For the eigenvalue $\lambda_1 = 1$ for example, the eigenvector $\xi^1 = [\,\xi_1^1\ \xi_2^1\,]^T$ is determined by the system

$$(A - 1I)\xi = \mathbf{0},$$

which becomes:

$$\begin{cases} 1.2\xi_1^1 & - & 0.4\xi_2^1 & = 0 \\ -2.4\xi_1^1 & + & 0.8\xi_2^1 & = 0 \end{cases}$$

It is easy to see that the two equations in this system are equivalent; the second equation can be obtained from the first by multiplication with a factor $m = -2$. This is to be expected and is an effect of the matrix of this system being singular. Thus the eigenvector has to obey only one equation, $1.2\xi_1^1 - 0.4\xi_2^1 = 0$, or $\xi_2^1 = 3\xi_1^1$. Stated otherwise, this means that all the points on the line with slope three through the origin are possible solutions to this equation, so the eigenvector is aligned with that line. Repeating the process with the second eigenvalue $\lambda_2 = 3$, one gets $\xi_2^2 = -2\xi_1^2$. This is the line that is plotted in figure 2.1, where the vector $\mathbf{x} = [\,-1\ 2\,]^T$ was chosen along this direction, resulting in its being stretched three times by the action of the matrix.

Let us turn now to the relationship of the eigenvalues and eigenvectors with matrix multiplication. For this purpose, it is useful to collect the two eigenvectors of A as columns of a single matrix that will be denoted by R, given by

$$R = \begin{pmatrix} 1 & -1 \\ 3 & 2 \end{pmatrix} = (\,\xi^1 \mid \xi^2\,)$$

By the definition of matrix multiplication, it can be verified that

$$AR = (\,A\xi^1 \mid A\xi^2\,) = \Lambda R$$

where the newly-introduced matrix Λ is a diagonal matrix that contains the eigenvalues in the same order as the corresponding eigenvector columns in R, i.e.

$$\Lambda = \begin{pmatrix} 1 & 0 \\ 0 & 3 \end{pmatrix}.$$

The same result in linear algebra cited above also guarantees that R has an inverse if the eigenvalues are different, so that we can write what is known as a *similarity transformation* between the original matrix A and the diagonal matrix Λ,

$$A = R\Lambda R^{-1}, \quad \Lambda = R^{-1}AR.$$

It therefore follows that multiplication of A by itself can be written in terms of powers of Λ:

$$A^2 = AA = \left(R\Lambda R^{-1}\right)\left(R\Lambda R^{-1}\right) = \left(R\Lambda^2 R^{-1}\right),$$

and similarly $A^k = \left(R\Lambda^k R^{-1}\right)$. This makes the powers of A much easier to calculate, because the powers of Λ, as can be easily checked, are given by:

$$\Lambda^k = \begin{pmatrix} \lambda_1^k & 0 \\ 0 & \lambda_2^k \end{pmatrix}.$$

Thus, however large k is, one only needs two matrix multiplications (by R and R^{-1}) to compute A^k. This result greatly simplifies the study of the convergence of iterative methods for linear systems, as will be seen in a subsequent chapter.

Computing the determinant of larger matrices can be reduced to computing determinants of smaller matrices using the *cofactor expansion* across a row or a column. For the matrix $A = (a_{ij})$ of order n, the expansion across a row has the form

$$\det(A) = \sum_{j=1}^{n}(-1)^{i+j}a_{ij}D_{ij},$$

where D_{ij} is the determinant of the order $(n-1)$ matrix that is obtained by removing line i and column j from the matrix A. For example, let

$$A = \begin{pmatrix} 3 & 1 & -1 \\ 2 & 2 & -1 \\ 2 & 2 & 0 \end{pmatrix}.$$

Then one can use the cofactor expansion across the first row of $A - \lambda I$ to obtain the characteristic equation for A:

$$
\begin{aligned}
\det(A - \lambda I) &= \begin{vmatrix} 3-\lambda & 1 & -1 \\ 2 & 2-\lambda & -1 \\ 2 & 2 & 0-\lambda \end{vmatrix} \\
&= (3-\lambda)\begin{vmatrix} 2-\lambda & -1 \\ 2 & 0-\lambda \end{vmatrix} - (1)\begin{vmatrix} 2 & -1 \\ 2 & 0-\lambda \end{vmatrix} \\
&\quad + (-1)\begin{vmatrix} 2 & 2-\lambda \\ 2 & 2 \end{vmatrix} \\
&= (3-\lambda)(\lambda^2 - 2\lambda + 2) - (-2\lambda + 2) - (2\lambda + 2) \\
&= \lambda^3 - 5\lambda^2 + 8\lambda - 4 \\
&= (\lambda - 1)(\lambda - 2)^2.
\end{aligned}
\tag{2.2}
$$

Thus the eigenvalues of A are $\lambda = 1$ and $\lambda = 2$. The eigenvector ξ associated with $\lambda = 1$ is

$$(A - 1I)\xi = \begin{pmatrix} 2 & 1 & -1 \\ 2 & 1 & -1 \\ 2 & 2 & -1 \end{pmatrix}\begin{pmatrix} \xi_1 \\ \xi_2 \\ \xi_3 \end{pmatrix} = \begin{pmatrix} 0 \\ 0 \\ 0 \end{pmatrix}.$$

Expanding the above linear system leads to:

$$2\xi_1 + \xi_2 - \xi_3 = 0,$$
$$2\xi_1 + \xi_2 - \xi_3 = 0,$$
$$2\xi_1 + 2\xi_2 - \xi_3 = 0.$$

Since the first and the second equations are identical (this is usually stated by saying that the matrix $A - 1I$ has rank two instead of three), there are three unknowns but only two distinct equations. Therefore, the system has infinitely many solutions. If one chooses $\xi_1 = 1$ for example, one can easily find $\xi_2 = 0$ and $\xi_3 = 2$. Therefore, an eigenvector associated with the eigenvalue $\lambda = 1$ is $(1, 0, 2)^T$.

Before closing this section, it is worth mentioning an important class of matrices of which use will be made later on: the *positive-definite* matrices. These are the matrices A for which:

$$(\mathbf{x}, A\mathbf{x}) > 0,$$

for all non-zero vectors $\mathbf{x}$. If the matrix A is Hermitian, the positive-definiteness condition can be restated in terms of its eigenvalues: such a matrix is positive-definite if all its eigenvalues are positive. It is useful to point out that, for matrices with all real entries, this last case leads to the important class of *symmetric positive-definite* (SPD) matrices, for which the condition above can be equivalently written as:

$$\mathbf{x}^T A \mathbf{x} > 0, \tag{2.3}$$

and which also satisfy $a_{ij} = a_{ji}$.

2.6 Operation Counts

The representation of real numbers with a finite number of bits on a computer may lead to errors due to round-off. A further source of errors is the performance of arithmetic operations (like multiplication) involving such numbers, known as *floating-point operations* or *FLOPs*, in particular if these numbers contain decimal fractional parts. The generation and subsequent propagation of round-off errors are studied in more advanced numerical analysis courses. Because of the introductory nature of this text, these topics will not be further pursued here.

Being aware of the number of FLOPs (the abbreviation has become so common that nowadays capitals are not commonly used any more, so we'll write *flops* instead) required for a more complex evaluation is oftentimes useful: it provides a rough estimate for the computing time associated with that evaluation. For example, let's say one needs to multiply a matrix A of order n with a column vector $\mathbf{x}$. To compute the first element y_1 of the product $\mathbf{y} = A\mathbf{x}$ one needs to first evaluate the product $a_{11}x_1$ then, upon its storage in computer memory, add to it the product $a_{12}x_2$ and so on. Therefore, only for the computation of y_1, one needs to perform n multiplications and $n - 1$ additions, thus $2n - 1$ flops are involved for the first element. Since the same amount of work is required for all the elements in $\mathbf{y}$, the total number of flops involved in evaluating the product of an order n matrix with a vector, or a *matvec* operation as it is commonly referred to, is $n(2n-1) = 2n^2 - n$. When n becomes very large, the n^2 term grows much faster than the first power of n and quickly becomes the dominant term in this expression. A simplification was traditionally made when counting flops: since on most computer architectures multiplications and divisions were much more time consuming than addition and subtraction, it was rather the rule to count only the first two operations and not the latter. Thus, a matrix-vector operation is usually thought of a being performed at a cost rougly proportional to n^2. Furthermore, using this result, it is easy to calculate the price of an order-n matrix-matrix multiplication AB; one can think about it as multiplication of the matrix A with each one of the n columns of the matrix B. There are n such columns, and for each column we need $2n^2 - n$ flops, hence the total cost is $n^2(2n-1)$, growing with n^3 if only the multiplications are counted.

Operation counts like the above give us an idea of the computer cost (or time) necessary for performing a given operation. The picture is somewhat complicated because of different other variables involved. For example, there may be differences in hardware which impact the cost of accessing computer memory or differences in software (a matrix-vector multiplication algorithm written in a powerful language like C may be much faster to execute than a Python code for the same purpose). Nevertheless, considering the matrix-vector product as an example, a realistic estimate of the computer cost would be of the form Cn^2 if only multiplications are counted, where C is some constant that depends on the particular implementation. The important question is, of course, what happens as n becomes larger. A very common notation is used in fields such as computer science and numerical analysis to convey this information, the *"big O"* notation. To indicate the fact that the cost of a matrix-vector product grows no faster than n^2, we say it is $\mathcal{O}(n^2)$, pronounced 'big oh of n squared'. In this notation, the constant C multiplying the n^2 term need not be identified exactly. More exactly, suppose a function $f(n)$ represents the number of basic operations needed to perform a computation. One says that $f(n)$ is $\mathcal{O}(g(n))$ as $n \to \infty$, where $g(n)$ is some other function, if the limit $\lim_{n\to\infty} f(n)/g(n)$ of the ratio of these two functions is finite and equal to a constant. For example, if $f(n) = 20n^4 + 51n^2 + n$, one can say $f(n)$ is $\mathcal{O}(n^4)$. Usually the equal sign is used to state this behavior of the function, expressed as $f(n) = \mathcal{O}(n^4)$. Notice the convenience of this notation, as it allows us to neglect the various constants in front of the various powers of n; also note that the use of the equal sign might be somewhat misleading, because not all $\mathcal{O}(n^4)$ functions have the same form as $f(n)$. The function $g(n)$ is known as a *gauge function* and is usually chosen among the functions with a more intuitive behavior such as the power, logarithmic and exponential functions.

The "big O" notation is also useful to qualify the decay of some function of a given argument as this argument goes to zero. For an example, consider the cosine function $\cos(x)$ which around $x = 0$ has the Taylor series expansion

$$\cos(x) = 1 - \frac{x^2}{2!} + \frac{x^4}{4!} - \frac{x^6}{6!} + \dots.$$

It is easy to notice that, as x becomes smaller and smaller, a pretty good first approximation would be $\cos(x) = 1$. This approximation of course neglects all the other terms, the most important of these terms being the one in the second power of x. It is usual in this case to write $\cos(x) = 1 + \mathcal{O}(x^2)$ as $x \to 0$ to indicate the terms that are dropped. Of course one can also write $\cos(x) = 1 - x^2/2 + \mathcal{O}(x^4)$ for a more accurate representation that keeps the first two terms in the Taylor series and shows that the error involved in this latter approximation decays as x^4. More exactly, a function $f(x)$ is $\mathcal{O}(g(x))$ as $x \to 0$ if the limit of the ratio $f(x)/g(x)$ as $x \to 0$ is a constant.

2.7   Exercises

Ex. 1 Compute the dot product $\mathbf{x}^T\mathbf{y}$ and the inner products $(\mathbf{x}, \mathbf{y})$ and $(\mathbf{x}, \mathbf{x})$ for the two vectors defined by:

$$\mathbf{x} = \begin{bmatrix} 1 - i \\ 2 \\ 3 + 2i \\ 4i \end{bmatrix}, \quad \mathbf{y} = \begin{bmatrix} 2i \\ -1 \\ 2 - 3i \\ 1 + i \end{bmatrix}.$$

What is the norm of the two vectors?

Ex. 2 Find the eigenvalues of the diagonal matrix D given by:

$$D = \begin{pmatrix} 2 & 0 \\ 0 & -5 \end{pmatrix}.$$

Generalize to the case of a diagonal matrix of order n. Furthermore, check that the following two vectors are eigenvectors corresponding to the two eigenvalues:

$$\xi^1 = \begin{bmatrix} 1 \\ 0 \end{bmatrix}, \quad \xi^2 = \begin{bmatrix} 0 \\ 2 \end{bmatrix},$$

and compute the matrix D^4. Finally, find the inverse matrix D^{-1}. You can easily do this by inspection.

Ex. 3 Calculate $A\mathbf{x}$ and AB for the quantities given below. Is the product BA equal to AB?

$$A = \begin{bmatrix} 2 & 1 & 2 \\ -1 & 1 & 3 \\ -2 & 3 & 5 \end{bmatrix} \; ; \; B = \begin{bmatrix} 1 & 2 & 1 \\ 3 & 7 & 2 \\ 3 & 3 & 5 \end{bmatrix} \; ; \; \mathbf{x} = \begin{bmatrix} 2 \\ 1 \\ 3 \end{bmatrix}.$$

Then compute the inner product and the *outer product* X of the vector $\mathbf{x}$ with itself. The latter is defined by $X = \mathbf{x}\mathbf{x}^T$.

Ex. 4 Consider the same matrices A and B as in the previous exercise, and let

$$C = \begin{bmatrix} 2 & 1 & 0 \\ 0 & 2 & 1 \\ 3 & 1 & 3 \end{bmatrix}.$$

Verify the following identities: $(A+B)C = AC + BC$, $(AB)C = A(BC)$ and $(A+B)+C = A+(B+C)$.

Ex. 5 Is the matrix A below a singular matrix? If yes, clearly state why; if not, find its inverse. Also find the eigenvalues of A.

$$A = \begin{bmatrix} 2 & 6 \\ 1 & 1 \end{bmatrix}.$$

Ex. 6 Solve the following system of three equations with three unknowns:

$$\begin{cases} 3x + y - z &= 3 \\ 2x + 5y + z &= 8 \\ x - y + 4z &= 4 \end{cases}$$

Ex. 7 Find the two factors L and U of the system matrix in the previous exercise.

Ex. 8 Find the eigenvalues and eigenvectors of the matrix:

$$A = \begin{bmatrix} 2 & 1 \\ 3 & 4 \end{bmatrix}.$$

Using this information and similarity transformations, calculate A^5.

Ex. 9 Find the eigenvalues and eigenvectors of the matrix:

$$A = \begin{bmatrix} 4 & 0 & 1 \\ -2 & 1 & 0 \\ -2 & 0 & 1 \end{bmatrix}.$$

Ex. 10 Show that the matrix A below is symmetric and positive-definite (SPD).

$$A = \begin{bmatrix} 2 & 1 \\ 1 & 3 \end{bmatrix}.$$

Chapter 3

Basics of MATLAB®

3.1 Defining and Using Scalar Variables

When starting the MATLAB environment for the first time, the graphical user interface (GUI) might look similar to what you see in figure 3.1. The topmost region contains various menus that are context-dependent. The *command window*, located in the middle bottom region, is where commands that are evaluated immediately can be introduced at the MATLAB prompt. The *file explorer*, the left column in this figure, shows the content of the current MATLAB working directory. Files in that directory can be opened in the editor by double-clicking their names in the file explorer window. The *MATLAB workspace*, which appears as the rightmost column here, shows all the data that have been defined in the current session. The *editor window*, located above the command window in this snapshot, will only be present if a file has been opened or if code, to be later saved in a file, is being written.

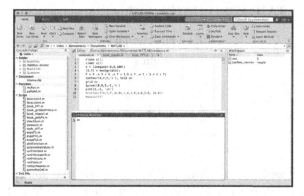

Figure 3.1: The different window panes in the MATLAB graphical user interface.

MATLAB can easily be used as a scientific calculator. For example, one can type the following in the command window when presented with the prompt (the two "greater than" signs):

```
>> 2 + 3 - ( 5 -1 )
```

and upon pressing the return key MATLAB returns the answer:

```
ans =
    1
```

The usual *infix notation* for addition and subtraction has been used here. MATLAB also allows the use of the less common *functional notation* for these basic operations; the following returns the same result

```
>> minus( plus(2,3), minus(5,1) )
```

Functional notation is very useful for operations that we cannot type because of the limited extent of available inputs on the keyboard. One may think about some of the available operations as functions and may write them as such in mathematical notation as well, hence the name. Notice, however, that in mathematics notation some may be represented by symbols, such as the square-root symbol, that does not appear on the keyboard. Therefore, it is absolutely natural to introduce such operations as named functions:

```
>> sqrt(5)
```

For multiplication and division of scalars we may also use the common asterisk and forward slash notation, but some care is required when working with vectors and matrices (section 3.4). The two notations for the most common arithmetic operators are summarized in table 3.1:

TABLE 3.1: MATLAB arithmetic operators

Infix	Functional
+	plus($\cdot$, $\cdot$)
-	minus($\cdot$, $\cdot$)
*	times($\cdot$, $\cdot$)
/	rdivide($\cdot$, $\cdot$)
^	power($\cdot$, $\cdot$)

In order to perform more involved computations, one must have access to places in computer memory for storing intermediary, as well as final, results. These are commonly identified, collectively, as the variables of the computation. While usually the programmer has a lot of freedom in choosing them, experience does make this choice clearer. To easily work with these places, instead of the tedious task of accessing computer memory locations, it has become common practice to refer to them via names that can be easily related to; these are the variable names. MATLAB is pretty flexible in its allowed range of variable names. It is both good practice and required in MATLAB to start a variable name with a letter, after which one may use any combination of letters, digits and the underscore character. Variable names in MATLAB are case-sensitive; thus, `myRatio` and `MyRatio` are two different variables. Unlike other languages, in MATLAB you don't need to declare a variable before first using it. Therefore, the first time one uses a variable is usually by assigning it a value. This can be done using the equal sign as below:

```
>> myRatio = 1/2
```

which displays:

```
myRatio =
        0.5
```

then returns the command prompt. Once a variable is assigned a value it can be used in further computations; reassigning a value to a variable is one such computation that occurs rather frequently. For example, with `myRatio` initialized as above, the command

```
>> myRatio = myRatio + 2
```

will return

```
myRatio =
     2.5
```

It is very important for people coming from a mathematics background to thoroughly contemplate the fact that the equal sign is not used to indicate equality, but assignment: the value on the right of the equal sign is assigned to the variable on the left. Therefore, it is imperative that whatever quantity appears on the right-hand side be already defined, so that the right-hand side can be readily evaluated. Had the quantity `myRatio` not already been initialized to the value one half, it would have made no sense to use it on the right-hand side of the above assignment.

Once a variable is defined, it can be checked against different values using relational operators that return a result which can be either true (represented by a one in MATLAB) or false (zero in MATLAB). A variable that can take values only in the set $\{F = 0, T = 1\}$ is commonly referred to as a boolean variable.

```
>> myRatio > 2
 ans =
     1
>> b = myRatio == 2  % b is a boolean
  b = 0
```

Table 3.2 lists the most common relational operators in both infix and functional notation and also includes the logical operators that can be used to construct complex statements out of boolean expressions.

Here is another example of use of these operators; notice the three assignments grouped together on the same line, allowed in MATLAB:

TABLE 3.2: Relational and Logical Operators

Infix	Functional	Operator
==	eq	equal to
~=	ne	not equal
>	gt	greater than
<	lt	less than
>=	ge	greater than or equal to
<=	le	less than or equal to
&	and	logical AND
|	or	logical OR
~	not	logical NOT

```
>> x = 3; y=5; z =25;
>> ( x < y ) | ( z < y )
 ans =  1
>> ( x<y ) & ( z<y )
 ans = 0
```

MATLAB also provides the so-called short-circuit operators && and || for the logical "and" and "or", respectively. When using ||, for example, to evaluate $E||F$, statement F is not evaluated if E is found to be true because the result will definitely be true in this case. One can safely use these short-circuit operators as long as the operands (i.e. E and F) are scalars; problems may arise if the operands are themselves arrays of boolean values. Whenever in doubt, use the standard form listed in table 3.2.

One doesn't always need to assign the output of a computation to a variable. If no variable is provided, the result is assigned to the reserved variable **ans**, as already seen above. For example:

```
>>   1/2
```

produces

```
ans =
      0.5
```

The value of **ans** changes with every command that returns an output value not assigned to a variable.

Functions are an essential concept in mathematics; in MATLAB the meaning of the word "function" is extended to include blocks of code that perform a well-defined task, usually part of a larger program. Nevertheless, the usual mathematical function of one or several variables can be easily defined in MATLAB. For example, to create a function defined by $f(x) = x + 1$ we can use the command

```
>>  f = @(x) x + 1;
```

which creates the respective rule. This type of functions are currently[1] known as *anonymous functions*. Once the rule is defined, one can apply it to a number of variables, as long as their value was previously defined. Such value-holding variables are usually known as arguments, not unlike usual mathematics language. The commands below exemplify the use of a couple of functions with scalar arguments. Note that, while it is common practice to use x, y, etc. for the name of the input variables in the definition of the function, these generic names can be chosen to suit our purpose.

```
>>  f = @(x) x^2 + 1;
>>  y = 2;
>>  f(y)
ans = 5
>>  g = @(x,t) x^2 + t;
>>  g(2,3)
ans = 7
```

Several variables that store useful quantities are predefined in MATLAB. These include `pi`, which stores the value $\pi = 3.1415\ldots$ as expected, `eps` which holds the floating-point relative machine accuracy, together with maximum and minimum values for different variable types and the symbol `NaN` which stands for "Not a Number" and is displayed whenever a computation (such as an attempted division by zero or even by a very small number) fails to produce a result. The imaginary unit is stored by default in two variables, $i = j = \sqrt{-1}$. It is easily overwritten, as programmers tend to use variables with similar names i, j as loop indices, so plan carefully if complex numbers are needed somewhere in your code. The following commands show several of these variables and what happens when overwriting them. The first formatting command instructs MATLAB to print all available digits instead of only the default four.

```
>>  format long
>>  pi
ans = 3.141592653589793
>>  eps
ans = 2.220446049250313e-16
>>  intmin
ans = (int32)  -2147483648
>>  intmax
ans = (int32) 2147483647
>>  format short
>>  3 + 2*i % complex number
```

---

[1]Previously MATLAB used a different syntax to create such simple functions, which were known as inline functions.

```
3.0000 + 2.0000i
>> i = 2; % redefining i
>> 3 + 2*i
ans = 7
```

While overwriting these variables by assigning to them different values is allowed, this is not a safe practice, as the above shows: one may easily forget and end up with puzzling results. Moreover, as in any programming language, there are reserved keywords (i.e. `if`, `else`) that one shouldn't use for variable names. Although not mandatory, tradition from the FORTRAN era, which has become common practice among programmers, prompts us to start the name of integer-valued variables with the letters `i, j, k, l, m, n`. Finally, if you end a command with a semicolon, the output of that command is suppressed as could be seen in some of the above examples.

3.2   Saving and Reloading the Workspace

Once you have defined a number of variables, you may need to save your work in case you want to restart at a later time. As mentioned before, the sum total of the variables that have been defined at any time is known in MATLAB as the workspace. While most users will prefer to save their code as a *script* file, a strategy that will be discussed later, one can also use the **Save** button in the menu or, equivalently, use the command prompt to save the workspace. For example,

```
>> save filename
```

will write **all** the current workspace variables in a *binary* file called `filename.mat`. This includes all scalar variables and anonymous functions, as well as arrays and settings for your figures, if there are any. Data in such binary files needs to be accessed with MATLAB; it can't be readily inspected with another program. To reload the data in a new session, one can use

```
>> load filename
```

with the same filename that was used to save the session. Try to only use `save` either when you want to close your MATLAB session in a rush and start from the same point at a later time. At all other times, you may find that it is a much more useful strategy to create and save script files.

One can also save the data, all or part of it, in plain text (ASCII) format. For example, to save a variable with the name x, one can use:

```
>> save x.dat  x  -ASCII
```

This will create a file with the name *x.dat* in the MATLAB working directory. This file can be explored with any program that handles text, such as Notepad. One may then reload the data at a later time, as needed:

>> load x.dat

With the speed of currently available computers, saving and reloading data in the workspace is usually necessary only when performing large computations that require very long times but can be conveniently broken up in to several segments. For the examples in this book, this will not be the case. Therefore, to reiterate, the preferred way to save your work will be to write scripts and function files which can then be reloaded and run with different input parameters.

3.3 Defining and Using Arrays

The previous chapter introduced vectors and matrices, two types of indexed collections that are most useful for performing repetitive computational tasks. Knowing how to define and operate on them is very important for MATLAB proficiency. While the scalar variables that have been discussed so far are in fact also treated by MATLAB as 1×1 matrices, arrays with more than one element need to be handled with care: one must distinguish carefully between collective and element-wise operations when such arrays are involved.

To construct vectors and matrices MATLAB uses the collection (square) brackets. *Row vectors* can be created with:

>> rVec = [1 2 3];

and are oriented horizontally. In this command, the elements in a row vector may be separated either by pressing the space bar or by a comma:

>> rVec = [1, 2, 3];

One of the most immediate tasks for which vectors are useful is plotting. Consider, for example, that you know the temperature for the past week. You could plot the temperature versus the day of the week as follows:

>> day = [1, 2, 3, 4, 5, 6, 7];
>> temp = [71, 72, 63, 64, 64, 70, 73];
>> plot(day, temp)

which will produce a MATLAB figure similar to figure 3.2 in a new figure window.

Column vectors, oriented vertically, are created with:

>> cVec = [1 ; 2 ; 3];

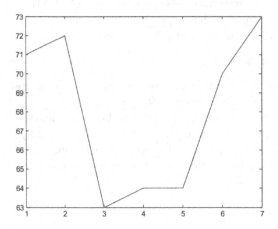

Figure 3.2: A simple figure produced using `plot` with no extra options.

The semicolons indicate advancement to the next row. Alternately, a carriage return can be used for the same purpose:

```
>> cVec = [1
            2
            3]
```

Both of the above commands create the column vector:

$$\text{cVec} = \begin{pmatrix} 1 \\ 2 \\ 3 \end{pmatrix}.$$

Vectors can also be created with the MATLAB-specific *colon operator* as in the following command:

```
>> ix = 0 : 1 : 5
```

which returns the row vector $\mathbf{ix} = \begin{pmatrix} 0 & 1 & 2 & 3 & 4 & 5 \end{pmatrix}$, while

```
>> iy = 0 : 2 : 6
```

creates $\mathbf{iy} = \begin{pmatrix} 0 & 2 & 4 & 6 \end{pmatrix}$. The functional notation for the same operation is:

```
>> iy = colon(0, 2, 6);
```

and it comes in handy when the infix notation is ambiguous. Even if both commands produce the same sequence, compare for example `colon(0, 2, 7+6)` to `0:2:7+6`. Related to the colon operator is the `linspace` command.

Although more seldom used, the latter has almost the same function: it generates a vector, with a specified number of elements this time, between two limits. The advantage of `linspace` is that one doesn't need to figure out the step. For example:

```
>> iz = linspace(0, 5, 4)
   iz =
      0   1.6667.   3.3333   5
```

The step between the successive entries produced by the colon operator doesn't need to be an integer;

```
>> ix = 0 : 0.1 : 1
```

gives $ix = \begin{pmatrix} 0 & 0.1 & 0.2 & 0.3 & \dots & 1 \end{pmatrix}$.

Thus, the basic structure of the colon operator is

start-value : step : end-value

or

colon(start-value,step,end-value)

while the `linspace` command has the form

linspace(start-value, end-value, numberOfPoints)

If the step is chosen such that the end-value is not reached exactly, then the last value is the one just below the end-value. For example:

```
>> x = 0 : 2 : 9
```

returns $x = \begin{pmatrix} 0 & 2 & 4 & 6 & 8 \end{pmatrix}$.

Finally, if the step is not specified, it is taken by default to be one;

```
>> x = 0 : 5
```

gives $x = \begin{pmatrix} 0 & 1 & 2 & 3 & 4 & 5 \end{pmatrix}$.

Matrices can also be generated easily by listing their elements row by row and separating the different rows with semicolons:

```
>> A1 = [1 2 3; 3 1 2; 3 2 1]
```

$$A1 = \begin{pmatrix} 1 & 2 & 3 \\ 3 & 1 & 2 \\ 3 & 2 & 1 \end{pmatrix},$$

```
>> A2 = [ 1:3 ; 2:4 ; 5 6 7 ]
```

$$A2 = \begin{pmatrix} 1 & 2 & 3 \\ 2 & 3 & 4 \\ 5 & 6 & 7 \end{pmatrix},$$

or by inputting separate rows with the return key, instead of semicolons. Once a matrix is defined, its dimensions can be accessed using the function `size`, as follows:

```
>> size(A1)
ans = 3 3
>> size(A1,1)
ans = 3  % number of rows in A1
```

MATLAB also conveniently offers several predefined matrices that occur very often; you can explore for example the help pages for **zeros**, **ones** and **eyes**. Furthermore, one can create a matrix whose entries are random numbers uniformly distributed between zero and one using the **rand** command. Finally, a matrix whose only non-zero elements are on the diagonal can be created using **diag**, which takes as argument the vector of diagonal values.

Accessing individual elements of vectors and matrices can be done with a command of the form `ObjectName(i,j)`, where i is the row and j the column of the wanted element. For example,

```
>> A2(2,3)

ans =

    4
```

while

```
>> x = A2(2,:)
```

will access all elements of the second row of A2:

$$x = \begin{pmatrix} 2 & 3 & 4 \end{pmatrix}.$$

Similarly,

```
>> x = A2(:,3)
```

will access all elements of the third column of A2:

$$x = \begin{pmatrix} 3 \\ 4 \\ 7 \end{pmatrix}.$$

The MATLAB colon operator helps us perform most manipulations and extractions of matrix data, as in the following examples. The very useful **end** keyword is a placeholder for the last row or column, as appropriate.

```
>> A2(2:3, 2)
```

$$\begin{pmatrix} 3 \\ 6 \end{pmatrix}$$

```
>> A2(2:end, 2)
```

$$\begin{pmatrix} 3 \\ 6 \end{pmatrix}$$

```
>>A2(2, 2:end)
```

$$\begin{pmatrix} 3 & 4 \end{pmatrix}$$

```
>>A2(1:end-1, 2:end)
```

$$\begin{pmatrix} 2 & 3 \\ 3 & 4 \end{pmatrix}.$$

3.4   Operations on Vectors and Matrices

Let's start by listing some elementary functions that can be used on vectors. The following return a scalar value, are predefined in MATLAB and do just what you expect them to do:

```
>> sum(rVec);
>> prod(rVec);
>> mean(rVec);
>> max(rVec);
>> min(rVec);
```

As an example of use, the function below takes a positive integer n and checks whether the sum of the first positive integers up to n stored in a vector created with the colon operator is indeed $n(n+1)/2$, returning true if that is the case.

```
>> checkSum = @(n) sum(1:n) == n * (n+1) / 2;
>> checkSum(5) % output : logical 1 (i.e. true)
```

Cumulative versions of the sum and product also exist and return vectors of the same size as the vector they operate on. For example, with rvec = [1 2 3], we get

```
>> cumsum(rVec)
ans = 1 3 6
>> cumprod(rVec)
ans = 1 2 6
```

Most of the usual mathematical functions work on vectors and matrices, as well as scalars. They take a vector or matrix as input and return an object of the same size, with the function applied to all the elements of the input. Examples of such functions include sin and all the other trigonometric functions, the exponential function exp and the logarithm log, etc.

Since MATLAB was designed for matrix manipulation, all its variables are considered matrices by default and the usual arithmetic operators also expect matrices as operands. Operands with scalar values are simply treated as matrices with only one row and one column; row vectors are obviously matrices with only one row, etc. Therefore, if we try to perform the following:

```
>> rVec * rVec
```

we will get an error message since this operation is illegal. Specific to MATLAB are the element-wise operators; element-wise variants for multiplication, division and raising to a power exist and are typed as the respective operator preceded by a dot. Thus, one can perform element-wise multiplication of two row vectors of the same size:

```
>> rVec .* rVec
ans = 1 4 9
>> rVec .^ 2
ans = 1 4 9
>> rVec ./ 2
ans = 0.5000 1.0000 1.5000
```

while `rVec ^ 2` and `rVec / 2` will produce error messages. Note however that adding a scalar value to a matrix is allowed: the scalar will be added to all matrix entries.

From the above, it follows that one needs to exercise care when writing functions that may have array inputs; the operations the function performs must be valid for the type of input that is presented to the function. Let us consider an example:

```
>> f = @(x) x + 1;
>> g = @(x) x^2 + 1;
>> x = 3;
>> v = [ 1, 2 ];
>> f(x)
ans = 4
>> f(v)
ans = [ 2 3]
>> g(x)
ans = 10
>> g(v)
```

In this example, the last command will produce an error message because v^2 is not defined. In order for the function $g(x)$ to admit a vector as input and produce a vector with the same number of elements, the code should be modified accordingly; similarly if the aim is to produce a scalar. The commands below show two such possibilities:

```
>> g1 = @(x) x.^2 + 1; %element-wise multiplication
                       %produces a vector
```

```
>> g2 = @(x) x*x' + 1; %uses dot product
                                    %produces a scalar
>> v = [ 1, 2 ];
>> g1(v)
ans = [ 2 5 ]
>> g2(v)
ans = 6
>> g3 = @(x) sqrt(x) * sin(x); %won't work with vectors
>> g4 = @(x) sqrt(x) .* sin(x); %works with vectors
```

If an anonymous function expression becomes too long to be typed on one line, one can use the ellipsis ..., which serves the purpose of a continuation mark:

```
>> newf = @(x) sin(x) - ...
                1.3*cos(1.45*x)
```

Many MATLAB users use the apostrophe, as shown for **g2** above, to *transpose* a matrix; in the example, the variable **x** is supposed to be a row vector and **x'** is a column vector. However, the apostrophe only produces the transpose for real-valued matrices, while for complex-valued matrices it performs the conjugate transpose (the adjoint). It is preferable to get used to the element-wise apostrophe if you prefer infix notation or, probably even better, to use the functional notation **transpose**. For the real-valued matrix **A2** as defined above

```
>> B1 = A2.';
>> B2 = transpose(A2);
>> B3 = A2';
```

all produce the same result,

$$B1 = B2 = B3 = \begin{pmatrix} 1 & 2 & 5 \\ 2 & 3 & 6 \\ 3 & 4 & 7 \end{pmatrix},$$

whereas:

```
>> x = [3 + 2*i, 1, 2]
>> y = x.'
y = [ 3+2i; 1; 2]
>> z = x'
z = [ 3 - 2i; 1; 2]
```

Note that 3 + 2i will also work to input the same complex number for the first element of **x**.

Relational (logical) operators working on arrays produce arrays of boolean values of the same size. For example:

```
>> x = [1 2 0 3 0 5]
>> x > 2
ans = 0 0 0 1 0 1
>> x == 2
ans = 0 1 0 0 0 0
```

The second command checks for equality of the elements of vector **x** to the value 2 (note the double equal sign). However, most of the time, one is interested in the index set of the values in an array that satisfy some condition. In MATLAB, it is easy to return the indices of the subset of elements, in an array of zeros and ones, that have the value one (true). This is accomplished using the **find** command which takes a boolean array as input:

```
>> x == 0

ans =

     0     0     1     0     1     0
>> find(x == 0)
ans =

     3     5
```

This compares elements of **x** to the specified value (zero in this case) and returns the index number of those that match (elements in positions three and five). Some further examples:

```
>> find(x > 2)

ans =

     4     6

>> find(x > 11)

ans =  [ ]   % the empty vector
```

The above is MATLAB notation for the *empty vector*. Furthermore,

```
>> A = [1 2 3; 4 5 6];
>> B = [2 3 3; 5 5 7];
>> find(A == B)
ans = 4 5
```

might raise some question marks, as the fourth and fifth elements of the matrices are reported equal. To understand these numbers, it is good to know that MATLAB orders the matrix elements in column-major order; that is, the first column followed by the second and so on. Thus, the matrix A in the above example is stored by MATLAB as the sequence of values $[1, 4, 2, 5, 3, 6]$.

The last element in a vector can be accessed in a few ways, probably most conveniently with the **end** index that has already been introduced before:

```
>> length(x)

ans =

     6

>> x( length(x) ) == x(6)

ans =

     1

>> x(end)

ans =

     5
```

Erroneous indices will return error messages. To extend (or add elements to) a vector, one can use:

```
>> x(end + 1) = 77
```

To prepend elements:

```
>> x = [11, 12, x]
```

Similarly, to append several elements at a time:

```
>> x = [x, 13, 14]
```

Finally, it is worth pointing out that MATLAB provides a convenient command to compute the eigenvalues and their associated eigenvectors for a given matrix. Consider again the 3×3 matrix in section 2.5

$$A = \begin{pmatrix} 3 & 1 & -1 \\ 2 & 2 & -1 \\ 2 & 2 & 0 \end{pmatrix}.$$

Recall that the eigenvalues of A are $\lambda = 1$ and $\lambda = 2$. For $\lambda = 1$, the associated eigenvector is $\xi = (1, 0, 2)^T$. Since the eigenvectors are defined by their direction, one can normalize ξ by dividing all its components by its norm. This produces the unitary eigenvector $\tilde{\xi} = (1/\sqrt{5}, 0, 2/\sqrt{5})^T = (0.447213595499958, 0, 0.894427190999916)^T$. The use of the MATLAB command **eig** for computing the eigenvalues and the associated eigenvectors is demonstrated below:

```
>> A=[3 1 -1; 2 2 -1; 2 2 0]
A =

     3     1    -1
     2     2    -1
     2     2     0

>> [V, D]=eig(A)
V =

  -0.408248296858222   0.408248284069505   0.447213595499958
  -0.408248296858221   0.408248284069503  -0.000000000000000
  -0.816496574533367   0.816496587322085   0.894427190999916


D =

   2.000000046988748                   0                   0
                   0   1.999999953011250                   0
                   0                   0   1.000000000000000
```

The command returns a matrix D with eigenvalues on the diagonal and a matrix V whose columns are the associated eigenvectors. In our example, the last entry in the diagonal of D is $\lambda = 1$ and the associated normalized eigenvector is the last column of V.

3.5   More on Plotting Functions of One Variable

As seen in section 3.3, one of the important uses of vectors is for plotting. Here we'll explore plotting functions of one variable in more detail. Right off the bat, let's remark that MATLAB plots, produced on the screen, may not look very well when printed in black and white. The reason for this is that the default color for the graph of a function on the screen is blue, which prints as a shade of gray on the printer and hence may lack clarity. The plot function is very versatile however, and offers its users a large array of options to help them control the quality and amount of information their plots display. Here is how one can modify the plot in section 3.3 to make the line black instead, while also marking the data points with a star:

```
>> day = [1, 2, 3, 4, 5, 6, 7];
>> temp = [71, 72, 63, 64, 64, 70, 73];
>> plot(day, temp,'k-*')
```

This form of **plot** uses optional arguments, included in between quotes, to indicate that the line color to be used is black. Figure 3.3 shows the output from this command.

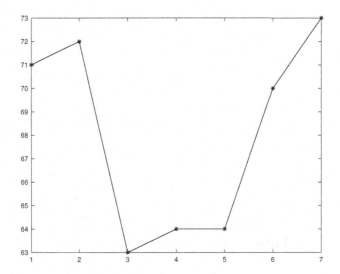

Figure 3.3: Temperature plot produced using **plot** with black line color and markers.

Let us now explore the various plotting options in more detail. In order to create a plot, the first thing we need to do, as we have already seen, is to define the functions to be plotted, i.e. vectors of coordinate pairs (x, y) in the case of functions of one variable. Most of the time this is easiest done using either the colon operator or the **linspace** command. The following creates a vector of equispaced points in the x direction, sometimes referred to as a mesh or a grid (arguably somewhat misleading, as the words rather invoke 2D connotations) in scientific computing. The values of the function $\exp(0.16x) * \sin(3x)$ are then computed for these points, and a basic plot using a red, thicker line is produced. Note the element-wise multiplication needed in the computation of $y1$.

```
>> x1 = 0 : 0.1 : 10;
>> y1 = exp(0.16*x1) .* sin(3*x1);
>> plot( x1, y1, 'r' , 'LineWidth',2)
```

This figure, just like the ones above, lacks most of the information that is useful on such plots: axis labels, a title and a legend describing the function being plotted. Such information can be added to the plot, as will be seen

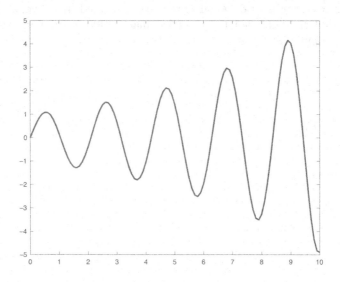

Figure 3.4: A plot with thicker, red line.

below. On a different track, it is also useful to remember that the points on the horizontal axis do not need to be equispaced. The following very coarse plot of the same function with a green line uses only a few points picked at random and shows again that, when plotting, MATLAB does nothing more than just joining successive points by a straight line. The illusion of a nice curve in the first plot is only due to the fact that the points are relatively closely spaced together.

```
>> x2 = [ 0 1 sqrt(2) pi 4 7.2 ] ;
>> y2 = exp(0.16*x2) .* sin(3*x2) ;
>> plot( x2, y2, 'g' )
```

Remember that the `linspace` command is similar to the colon operator, but allows you to specify the length of the vector, not the step, and is therefore also very useful in creating data sets to be plotted:

```
>> x3 = linspace(0, 10, 45);
>> y3 = exp(0.16*x3) .* sin(3*x3)   ;
>> plot( x3, y3, 'm--o');
```

The plot uses now a magenta dashed line with symbols (circles) at the data points. More information about the various options available with `plot` is available through its help page. Some of the most important of these are further explored below.

One can graph several data sets in a single plot, and there are at least two ways to do so. The first way is to use the `plot` command with all the three data sets listed one after the other, their particular options included. The following lines show how this can be done for the data defined above; in the corresponding figure 3.5 the first data set is plotted with the default line style, the second with asterisk markers and the third with a dotted line and square markers. The `figure` command creates a new figure handle, which can then be accessed via its integer identifier. Also exemplified is the use of `title` to assign a title to the plot, and `xlabel` and `ylabel` for labeling the axes.

```
>> figure(1)
>> plot(x1, y1, x2, y2, '-*', ...
              x3, y3, ':s','MarkerSize',9)
>> title('Three data sets')
>> xlabel( 'x' )
>> ylabel( 'Growing oscillation' )
```

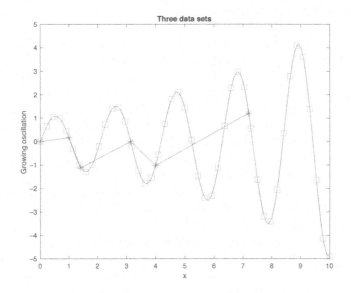

Figure 3.5: A plot of three data sets with various options.

A second way that MATLAB offers to plot several data sets is very useful when the particular data sets involved are obtained at different points in time, for example in a sequence of computations known as a loop. In this case one can initialize the respective figure and plot the various data sets as they become available, while keeping the pertaining figure active. Here is a very simple example of how this can be done:

```
>> figure(2)
>> plot( x1, y1, 'g' ), hold on
>> plot( x2, y2, 'r' )
>> plot( x3, y3, 'm' )
>> hold off
```

The `figure(2)` command again creates a new figure handle (now identified as number two); the following plotting operations will apply to this figure until we activate a different one using the same command with a different identifier. The new `hold on` command is necessary only after the first plot. All subsequent plots will be graphed in the same window, until the corresponding `hold off` is issued. Finally, a very useful capability of `plot` is that it can show legends for the various data sets involved. The location of the legend can be determined by the user, or can be left to MATLAB as shown below:

```
>> figure(3)
>> plot( x1, y1, x2, y2 )
>>legend('Set 1', 'Set 2',...
'Location', 'Best')
```

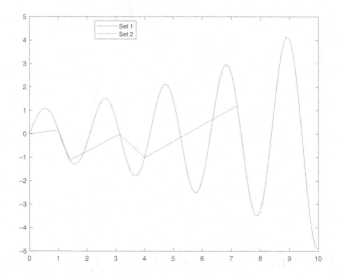

Figure 3.6: Two data sets with accompanying legend positioned by MAT-
LAB.

3.5.1 Subplots and Printing

The `subplot` command allows multiple graphs to be placed together in a tabular format. For example, to plot three data sets in three separate plot frames, one can create a plot with three rows and one column:

```
>> figure(4)                          %the 4th figure
>> subplot(3,1,1), plot(x1,y1), axis[-10 10 -10 10]
>> subplot(3,1,2), plot(x2, y2)
>> subplot(3,1,3), plot(x3, y3)
```

The `subplot` command has the general form:

```
subplot( row, column, graph_number )
```

The `axis` command has also been used to specify ranges of x and y:

```
axis[ xmin  xmax  ymin  ymax ]
```

For each subplot, one can still use the `xlabel`, `ylabel`, `title` commands. The reader should take a moment here to look for help on legend positions (i.e. `'North'`, `'Best'`, `'BestOutside'`) with `help legend`. One can also create legends containing numbers dependent on previous computation steps by converting them to strings using the `num2str` command. An example:

```
>> z = 2.1;
>> title([ 'Value of z:' num2str(z)])
```

As we saw, the [ ] delimiters create a vector of characters, which is then filled with the string 'Value of z:', after which the string containing the digits for the value of z is appended. One can also superpose a grid onto the plots with » `grid on` and remove it with the corresponding » `grid off`.

To print a figure to a file, one may conveniently use the figure menu or equivalently use the `print` command. This allows for a number of print formats (i.e. JPEG, EPS, etc.). Use `help print` for more information. For example, using

```
>> print -deps fig.eps
```

will print the current figure as an EPS (Encapsulated PostScript) file, a very useful vector format for a figure: it can be stretched or shrunk without loss of quality and is one of the preferred ways to insert figures in LaTex documents, alongside the portable document format (PDF).

3.6 Loops and Logical Operators

A classical MATLAB example concerns the famous Fibonacci sequence. You may be surprised to learn how often this sequence, in particular the Golden

Ratio associated with it, is encountered everywhere: in nature, arts, finance, let alone mathematics. The Fibonacci sequence is:

$$0, 1, 1, 2, 3, 5, 8, \ldots$$

and is obtained by starting with two integers (usually zero and one), then obtaining the next element in the sequence by summing up the previous two.

This sequence can be computed and stored within a vector which can be created element by element:

```
>> f(1) = 0;
>> f(2) = 1;
>> f(3) = f(1) + f(2);
>> f(4) = f(2) + f(3);
```

However, this process soon becomes tedious; it seems normal that one should be able to instruct the computer to save us from so much typing. In order to tell the computer to do the sum of the two previous values repeatedly, one can use a `for` loop. This is a basic command structure in MATLAB which in this case takes the form:

```
>> f(1) = 0;
>> f(2) = 1;
>> for i = 3:10
        f(i) = f(i-1) + f(i-2)
   end
```

The variable `i` is the *loop index*, a dummy variable that MATLAB uses to control the number of times it goes through a loop. The loop index will take all values from three to ten, inclusively, in this example. The computer will compute in turn $f(3), \ldots, f(10)$.

Since they allow us to indicate repetitive tasks to computers, loops are arguably the most important control structures in programming. Basic `for` loop commands may often be used to:

- compute the **sum** of all elements in a vector **x**. After initializing **x**, the sum can be computed as:

```
>> s = 0;
>> for i = 1:length(x)
        s = s + x(i)
   end
```

In mathematical notation, this is $s = \sum_{i=1}^{N} x_i$, where $N = \text{length}(x)$. Note that this sum can also be computed with the built-in MATLAB command **sum**, which we already saw. To make a difference between the fully spelled-out sum code above and **sum** when necessary, it is common to say that the full `for` loop code is explicit, versus the implicit version in **sum**.

- compute the **product** of all elements in a vector **x**.

After initializing **x**, the product is computed as:

```
>> p = 1;
>> for i = 1:length(x)
        p = p * x(i)
    end
```

In mathematical notation, this is $p = \prod_{i=1}^{N} x_i$, where $N = \text{length}(x)$.

The `for` loop can have a different step, obtained by the use of the colon operator. It can also go through indices specified in a vector. The examples below use this idea to compute two particular sums:

```
>> s = 0;               >> s = 0;
>> for j = 1:2:7        >> for j = [1 2 7]
        s = s + j               s = s + j
    end                     end
```

One of the many ways loops are useful is to obtain access to all the entries in a matrix. Because the matrix is a two-dimensional array, a couple of nested loops are needed in this case. In the example below, all one hundred matrix entries are set to two. Nesting loops may lead to large amounts of computation, so it is important to pay close attention to the way such computations are organized for efficiency. The two loops in the example are in the most natural but not the best configuration for MATLAB; more about this in section 3.7.

```
>> for i = 1:10
        for j = 1:10
            A(i,j) = 2;
        end
    end
```

While the `for` loop will be the preferred repetitive control structure in this book, there is another choice that may seem more appropriate under some circumstances. The MATLAB `while` loop checks that a condition is obeyed before going through the commands that make up the loop body and has the general form:

```
while condition
        commands
    end
```

For both loops, the `break` command inside a loop causes the computer to skip the remainder of the loop body and totally exit the loop. The related `continue` command only skips the remainder of the loop body for the current value of the loop index and restarts the loop with the next value of the index.

The `if` statement is another important program flow control structure, which offers a way for the program to proceed along different paths when needed. The basic format of an `if` statement is:

```
if (logical statement 1)
        (things to do 1)
    elseif (logical statement 2)
        (things to do 2)
    elseif (logical statement 3)
        (things to do 3)
                  .
                  .
                  .

    else
            (default things to do if the above fails)
end
```

The (`logical statement`) expressions used in the above `if` control structure are evaluated by MATLAB in order to decide about the execution of the corresponding block of code following them. They need to return a **boolean** value; remember that this means it can be thought of as either True/False or 1/0. The relational and logical operators in table 3.2 can be used to construct these expressions to the required degree of complexity.

A very closely related control structure is the `switch` statement. Its general form, usually somewhat easier to read than the corresponding `if` sequence, is:

```
switch (expression)
        case case1
                (things to do 1)
        case case2
                (things to do 2)
          .
          .
          .

        otherwise
                (things to do if all cases fail)
    end
```

A possible example of using the `switch` command to decide on a fictive schedule is shown below:

```
switch dayOfWeek
        case 1
                name = "Sunday";
        case {2,3,4,5,6}
                name = "Workday"
            otherwise
                name = "Saturday"
    end
```

Let us emphasize again that logical operators are used to construct possibly complex logical expressions, or statements, that can be evaluated as either true or false and used to control the flow of a program. This will gradually become clearer as our treatment proceeds.

3.7   Working with Indices and Arrays

It is useful at this point to make a closer connection between usual mathematical notation, which is oftentimes very clear but terse, and a more computer-friendly equivalent that we will call pseudo-code and which can more readily be translated into actual computer code. Let us start by looking at multiplication of a matrix A of order n with a column n-vector $\mathbf{b}$ to create a vector $\mathbf{c}$. Forget for a moment that MATLAB offers a simple way to perform this operation and let's try instead to dig a little deeper; this will greatly facilitate our understanding of the more complex tasks to be encountered later on. The formula for the generic element of the vector $[c]$, $c_i = \sum_{i=1}^{n} a_{ij} b_j$, encodes in fact all the information that is necessary for creating a computer code that performs the multiplication. Notice first that c_i needs to be computed for all the values of $i = 1, 2, \ldots, n$. Also remember that computing a sum is usually done with a loop, as we saw earlier. Hence one could expand on the terse mathematical formula to obtain the following write-up, usually known as a pseudo-code:

```
for all i = 1, 2,...,n
          set sum to zero
          for all j = 1, 2,...,n
                  add a(i,j)*b(j) to sum
          end
          set c_i to sum
   end
```

Being able to write pseudo-code is an important step in learning computer programming and computation and will accelerate the process. Take some time training yourself to do so. You can start, for example, by looking at the next more complicated operation, a matrix-matrix multiplication. Lay out the process in plain English, then change to more concise pseudo-code and count how many loops will be necessary in that case. Do this every time you have even a small amount of doubt about your code layout; sooner or later, you'll find that a clear picture emerges just by looking at the usual mathematical description of a problem.

When matrices become large, an important aspect that we need to be aware of is the fact that the way data is accessed in computer memory may have a large impact on computation efficiency. The entries in a matrix are stored in main memory in a linear fashion. In MATLAB, as opposed to other languages like C for example, the entries are arranged in a column-major order:

$$a_{11}, a_{21}, a_{31}, \ldots, a_{n1}, a_{12}, a_{22}, \ldots, a_{n2}, \ldots, a_{nn}.$$

Usually, when a generic entry a_{ij} is used in a computation, it is fetched from main memory into other memory locations that are much faster to work on (the cache). At the same time, an implicit assumption is made that close-by memory locations will also be used within a short time, so in order to speed up the computation several adjacent main memory locations are loaded into cache instead of only one. Thus, if one uses a_{11}, it is very likely that a_{21} and a_{31}, say, are also loaded in fast-access memory. It makes therefore sense in MATLAB code to go through the array in this column-major order instead of the (more intuitive probably) row-major order. Hence, while this is difficult to notice for small matrix sizes, expect the following MATLAB code:

```
for j = 1:n
        for i = 1:n
                A(i,j) = ...
        end
end
```

to perform faster than in the case when the i-loop is the outer loop.

3.8   Organizing Your Outputs

By default, MATLAB prints floating-point numbers with only four digits to the right of the decimal point. This can be controlled by the `format` command:

```
>> a = 6.626 e-34
>> format long
>> a
>> format short
>> a
>> b = 6.26267
>> format long
>> b
>> format short g
>> b
```

The "g" option specified for the last formatting command instructs MATLAB to use a general notation, with no exponent. Other options are available and left for the reader to explore using the MATLAB documentation.

Matrix output with MATLAB via the usual mechanism of just invoking the name of the variable to be displayed is usually acceptable, but there are instances when one may need more control over the output. The code excerpt below is a template that can be used to display the entries in a matrix with the desired precision and spacing. The code goes through the order-n matrix A line by line (the '`i`' index). For every line, it starts by printing an indenting space, then prints entries one by one in usual decimal point format, using nine characters altogether, six of which are for the decimal part. Each entry except the one in the last column is followed by three white spaces. At the end of a line, immediately after the last entry is printed, a new line (represented by a backslash immediately followed by the character **n**) is printed to jump to the next line. Again, for more information about the `fprintf` function and the different formatting options, the reader is invited to consult the MATLAB help pages.

```
for i = 1:n
        fprintf(' ')
        for j = 1:(n-1)
                fprintf('%9.6f   ', A(i,j) )
        end
        fprintf('%9.6f\n', A(i,n) )
end
```

It is oftentimes useful to keep track of the time that some computation requires, for example to confirm a previous analysis of an operation count. In MATLAB, this can be done easily with

```
>> tic
 "do the computation"
>> toc
```

The `tic` command starts an internal clock, which is stopped upon encountering the `toc` command; the latter also displays the elapsed time in between the two in seconds. Alternately, one can use the following sequence of commands:

```
>> a = now;
"do the computation"
>> b = now;
>> a                    % not a nice format
>> datestr(a)           % a format we can read
>> (b - a)*24*3600      % gives elapsed time in seconds
```

Other tools that might be useful, both in organizing the output of a piece of code and in debugging such code, are the built-in `disp` function and the `input` function. The first can output either a string of characters or the value of a variable then moves to a new line; the latter prints an optional message, then waits for input from the user. Their use is exemplified below:

```
>> disp ("The value of pi is:"), disp (pi)
>> help disp
>> input ("Pick a number: ")
>> help input
```

3.9 Number Representation

As is well known, a computer represents a real number by a sequence of binary bits. A binary bit has one of two values, either 0 or 1. Because of the finite number of bits used to represent both integer and floating-point numbers, values stored in MATLAB variables may be different from what an unexperienced user may expect. The translation between a series of zeroes and ones and the way we commonly write down numbers, referred to here as the *number representation*, is usually done using industry-wide standards. The most common representation used nowadays for real numbers is the IEEE standard 745 floating point representation. For what is known as single-precision arithmetic, this standard uses thirty-two binary bits to store a real number. When more precision is desired, numbers can be stored in double-precision. Double-precision has rather become mainstream; nevertheless, while it uses twice the amount of bits as the single-precision representation (so a total of sixty-four bits), the same concepts are at play.

As an example of how the representation works, suppose one wants to store the real number $x = 83.3$ on a computer using the IEEE 745 standard for single-precision arithmetic. The integer 83 is represented, in terms of the binary number system[2], by

$$83 = a_0 2^0 + a_1 2^1 + a_2 2^2 + a_3 2^3 + \dots,$$

where $a_0, a_1, a_2, \dots$ are either 0 or 1. To find $a_0, a_1, a_2, \dots$, one computes

$$83 \div 2 = 41 \quad \text{remainder} = 1 = a_0,$$
$$41 \div 2 = 20 \quad \text{remainder} = 1 = a_1,$$
$$20 \div 2 = 10 \quad \text{remainder} = 0 = a_2,$$
$$10 \div 2 = 5 \quad \text{remainder} = 0 = a_3,$$
$$5 \div 2 = 2 \quad \text{remainder} = 1 = a_4,$$
$$2 \div 2 = 1 \quad \text{remainder} = 0 = a_5,$$
$$1 \div 2 = 0 \quad \text{remainder} = 1 = a_6.$$

The above calculation indicates that integral part of the number, in this case 83, is represented by the following sequence of binary bits (notice that we commonly write the above binary coefficients from left to right, in the opposite order from the way they have been obtained, which is not necessarily how they are aligned in computer memory)

2^6	2^5	2^4	2^3	2^2	2^1	2^0
1	0	1	0	0	1	1

---

[2]A representation using only two digits, in this case 0 and 1, is known as a binary or base-2 representation, as opposed to the common base-10 representation which used the ten digits 0 through 9. The base is usually indicated by a subscript.

The binary representation for the decimal 0.3 is obtained using the negative powers of two,

$$0.3 = a_{-1}2^{-1} + a_{-2}2^{-2} + a_{-3}2^{-3} + \ldots,$$

where $a_{-1}, a_{-2}, a_{-3}, \ldots$ are again either 0 or 1. To find $a_{-1}, a_{-2}, a_{-3}, \ldots$, one computes

$$0.3 \times 2 = 0.6 \quad \text{the number of 1's is 0,} \quad a_{-1} = 0,$$
$$0.6 \times 2 = 1.2 \quad \text{the number of 1's is 1,} \quad a_{-2} = 1,$$
$$0.2 \times 2 = 0.4 \quad \text{the number of 1's is 0,} \quad a_{-3} = 0,$$
$$0.4 \times 2 = 0.8 \quad \text{the number of 1's is 0,} \quad a_{-4} = 0,$$
$$0.8 \times 2 = 1.6 \quad \text{the number of 1's is 1,} \quad a_{-5} = 1,$$
$$0.6 \times 2 = 1.2 \quad \text{the number of 1's is 0,} \quad a_{-6} = 1,$$

$$\vdots$$

At this point it is easy to see that a_{-6} to a_{-9} will repeat the pattern from a_{-2} to a_{-5}, which is $\boxed{1\,0\,0\,1}$. Similarly, a_{-10} to a_{-13} will repeat the same pattern, etc. We, therefore, represent 0.3 by the following sequence of binary bits

2^{-1}	2^{-2}	2^{-3}	2^{-4}	2^{-5}	2^{-6}	2^{-7}	2^{-8}	2^{-9}	2^{-10}	2^{-11}	2^{-12}	2^{-13}	
0	1	0	0	1	1	0	0	1	1	0	0	1	$\cdots$

or, using a common notation for the repeating pattern, $0.3 = (0.0\overline{1001})_2$. Therefore, putting together the parts before and after the decimal point, the floating-point binary representation for $x = 83.3$ can be seen to be:

$$83.3 = (1010011.0\overline{1001})_2 = (1.01001101\overline{1001})_2 \times 2^6$$
$$= (1.\underbrace{01001101\overline{1001}}_{\text{mantissa}})_\beta \times \beta^e, \tag{3.1}$$

where $(0.01001101\overline{1001})_\beta$ is the (normalized) mantissa (sometimes referred to as the fractional part of the logarithm, or simply the fractional part), $\beta = 2$ is the base, and $e = 6$ is the exponent. The way the mantissa is stored on a computer governs the accuracy, while the exponent gives the range, of the representation. The single-precision IEEE 745 floating point standard stores the mantissa and a shifted, or biased, exponent $b = e + 127$ using the following layout

sign	biased exponent	mantissa
1 bit	8 bits	23 bits

For positive numbers, the sign bit is 0; for negative ones, the sign bit is 1.

The number 0 is usually encoded as $+0$, but can be represented by either $+0$ or -0. The IEEE 754 floating point standard requires both $+0$ and -0. The reason to store a biased exponent is because 8 binary bits can represent at most 256 numbers, which can thus include 0 and all the positive integers from 1 to 255; in order to represent negative exponents without using a binary bit for its sign, the actual exponent $-127 \leq e \leq 128$ is represented by the biased exponent $0 \leq b = e + 127 \leq 255$ to bring it into this range; the positive shift, or bias, is thus 127. In the case of a double-precision floating-point number representation, the exponent uses eleven bits while the mantissa has fifty-two bits; in this case, the actual exponent range is $-1022 \leq e \leq 1023$, hence the bias is 1022. The smallest double-precision number that can be represented on the computer is therefore $2^{-1022} \approx 2.225073858507201e\text{-}308$. If someone tries to use a number smaller than this number, an error is obtained; this situation is known as double-precision floating point *underflow*. On the other hand, the largest double-precision number for MATLAB is $(2 - 2^{-52}) \times 2^{1023} \approx 1.797693134862316e\text{+}308$. A number bigger than this leads to a condition known as double-precision floating point *overflow*. We can easily find the MATLAB default smallest and largest numbers in their common base-ten representation:

```
>> format long
>> realmin
ans = 2.225073858507201e-308
>> realmax
ans = 1.797693134862316e+308
```

With this background, one can see that the single-precision IEEE standard 745 floating point representation for the real number $x = 83.3$, Eq. (3.1), is

0	1	0	0	0	0	1	0	1	0	1	0	0	1	1	0	1	0	0	1	1	0	0	1	1	0	0	1	1	0	0	1

Notice that the biased exponent is $b = 6 + 127 = 133$ in this case. Let us use the notation $fl(x)$ to denote the resulting representation of the real number x. Thus, for the single-precision IEEE 745 floating point standard, $x = 83.3$ can be represented in the computer as the following series of bits:

$$fl(83.3) = (1.01001101001100110011001)_2 \times 2^6. \qquad (3.2)$$

This representation is derived by discarding the infinite tail, an approach known as *chopping*

$$(0.\overline{1001})_2 \times 2^{-23} \times 2^6 = (0.0\overline{1001})_2 \times 2^{-22} \times 2^6 = 0.3 \times 2^{-16}$$

in (3.1), which means $fl(83.3) = 83.3 - 0.3 \times 2^{-16}$. The *round-off error* is $83.3 - fl(83.3) = 0.3 \times 2^{-16}$. Round-off errors are usually unavoidable in numerical computation, because it is impossible to represent most real numbers exactly in a computer. Due to round-off errors, even though the range for double-precision floating point numbers is between $2.225073858507201e\text{-}308$ and

1.797693134862316e+308, the accuracy of numerical computation is closely related to a quantity known as *machine epsilon*, which is discussed in the next section.

3.10   Machine Epsilon

Suppose the round-off error of a real number x is proportional to $|x|$, with a multiplication coefficient ϵ. The machine epsilon, denoted here by ϵ_{mach}, is the infimum (the smallest value) of the coefficient ϵ that still obeys the relation

$$|x - fl(x)| \leq \epsilon |x|. \tag{3.3}$$

The above definition is equivalent to

$$fl(x) = x(1 + \delta), \quad \text{with } |\delta| < \epsilon_{mach}. \tag{3.4}$$

From (3.3) one can see that this quantity has the meaning of a relative error,

$$\frac{|x - fl(x)|}{|x|} \leq \epsilon_{mach}. \tag{3.5}$$

The value of the machine epsilon depends on the rounding rules used in approximating $|x - fl(x)|$ in (3.5). Two possibilities are immediately obvious: either discard the digits that don't fit (the chopping technique already introduced earlier), or use the closest value that can be represented on the computer. The latter approximation technique is commonly referred to as *rounding* or, when one wants to be more precise, rounding-by-nearest. The corresponding values for machine accuracy are:

1. if $|x - fl(x)|$ is chopped, $\epsilon_{mach} = \beta^{1-p}$;

2. if $|x - fl(x)|$ is rounded-by-nearest, $\epsilon_{mach} = \frac{1}{2}\beta^{1-p}$.

Let us derive the quantity ϵ_{mach} for the case when the representation is chopped. Suppose a general representation of a floating-point number $fl(x)$ is

$$\pm(d_0.d_1d_2d_3 \ldots d_{p-1})_\beta \times \beta^e = \pm \left(d_0 + d_1\beta^{-1} + \cdots + d_{p-1}\beta^{-(p-1)} \right) \beta^e, \tag{3.6}$$

where $d_0 = 1$ and $d_i = 0$ or 1 for $i \geq 1$. For (3.2), $p = 24$, $\beta = 2$, and $e = 6$. Without loss of generality, one can assume $x > 0$; therefore:

$$\frac{|x - fl(x)|}{|x|} = \frac{|(d_0.d_1d_2d_3 \ldots d_{p-1}d_pd_{p+1} \cdots - d_0.d_1d_2d_3 \ldots d_{p-1})_\beta \times \beta^e|}{|(d_0.d_1d_2d_3 \ldots d_{p-1}d_pd_{p+1} \ldots)_\beta \times \beta^e|}$$

$$= \frac{|(d_p.d_{p+1} \cdots)_\beta \times \beta^{e-p}|}{|(d_0.d_1d_2d_3 \ldots d_{p-1}d_pd_{p+1} \ldots)_\beta \times \beta^e|}$$

$$= \frac{|(d_p.d_{p+1} \cdots)_\beta|}{|(d_0.d_1d_2d_3 \ldots d_{p-1}d_pd_{p+1} \ldots)_\beta|} \times \beta^{-p}.$$

The maximum of the above expression occurs when the fractional number has a minimum denominator and a maximum numerator. Notice that since $d_0 = 1 \neq 0$ the minimum value of the denominator occurs when $d_1 = d_2 = d_3 = \cdots = 0$. The minimum value of the denominator is 1. The maximum value of the numerator occurs when $d_p = d_{p+1} = d_{p+2} = \cdots = \beta - 1$, which suggests that the maximum value of the numerator is the sum of a geometric series

$$(\beta - 1) + (\beta - 1) \times \frac{1}{\beta} + (\beta - 1) \times \frac{1}{\beta^2} + \cdots = (\beta - 1)\left(1 + \frac{1}{\beta} + \frac{1}{\beta^2} + \cdots\right),$$

a value which is bounded above by β. Thus

$$\frac{|x - fl(x)|}{|x|} \leq \beta \times \beta^{-p} = \beta^{1-p} = \epsilon_{mach}.$$

The derivation for the round-to-nearest approach is similar, except one needs to consider two cases: (i) $d_p < \beta/2$ and (ii) $d_p \geq \beta/2$.

For the double-precision IEEE 745 floating point standard, $p = 53$ and the machine epsilon for chopping is thus $\epsilon_{mach} = 2^{-52} \approx 2.220446049250313e{-}16$. For the single-precision case, $p = 24$ and $\epsilon_{mach} = 2^{-23} \approx 1.1920929e{-}07$.

For MATLAB default precision, which is double precision, one can find ϵ_{mach}, letting $|x| = 1$ in definition (3.3), by invoking it via a built-in command:

```
>> format long
>> eps
ans = 2.220446049250313e-16
>> eps(1)
ans = 2.220446049250313e-16
```

Similarly, one can find ϵ_{mach} for MATLAB single-precision arithmetic by

```
>> format long
>> eps(single(1))
ans =   single
          1.1920929e-07
```

From definition (3.4), one can see that machine epsilon is the distance from the number 1.0 to the next largest floating point number, which can be written as:

$$fl(1 + \epsilon) > 1.$$

Thus, $fl(1 + \delta) = fl(1) = 1$, whenever $|\delta| < \epsilon_{mach}$. This can be confirmed by the following test:

```
>> format long
>> a = 1;
>> b = a + 1.5*10^(-16); % 1.5e-16<eps(1)=2.220446049250313e-16
>> b - a % an error occurs around a=1
ans = 2.220446049250313e-16 % this is eps
```

However, if we print eps(0) and let a=0 in the above test , we have

```
>> eps(0)
ans = 4.940656458412465e-324
>> a = 0;
>> b = a + 1.5*10^(-16); % 1.5e-16>eps(0)=4.940656458412465e-324
>> b - a % no error if a = 0
ans = 1.500000000000000e-16
```

The pre-defined MATLAB `eps` constant is indeed the distance from the number 1.0 to the next largest double-precision floating point number, which is 2^{-52}. The `eps(x)` function allows us to investigate the machine precision around some particular value x, defined as the distance from `abs(x)` to the floating-point number that is immediately larger.

MATLAB is very good at shielding the user from accuracy issues related to limited precision, which sometimes may be very surprising for a mathematician's mind. The built-in functions for variable-precision arithmetic, `vpa` and `digits`, from the Symbolic Math Toolbox, allow the user to control the number of digits used in the representation of a floating-point number. It may be argued that other languages are not that good at hiding representation problems, although sometimes this may actually be the more desirable outcome. As an example, let's take a look at distributivity of multiplication; the output below is directly copied from Python.

```
>>> 100*0.1 + 100*0.2
30.0
>>> 100*(0.1+0.2)
30.000000000000004
```

It certainly looks like MATLAB does give us the correct answer for the same calculation if we only use the default accuracy of sixteen digits and the default output of four decimal places:

```
>> 100*(0.1 + 0.2)
ans = 30.0000
>> 100*0.1 + 100 * 0.2
ans = 30
```

Here is the more illuminating computation however, first with more digits displayed for the same default accuracy, then with two extra digits of accuracy:

```
>> format long
>> 100*(0.1 + 0.2)
ans = 30.000000000000004
>> digits(18)
>> vpa( 100*(0.1 + 0.2) )
ans = 30.000000000000036
```

3.11   Exercises

Ex. 1 Create an anonymous MATLAB function that takes a set of numerical values, stored as a vector, and returns the boolean value `true` if the sum of all the values is larger than one hundred and `false` otherwise.

Ex. 2 Building on the previous function, define a new function that returns `true` if the sum of all the values is larger than some integer M but smaller or equal to a second integer N. The two integers M and N should be inputs to this function.

Ex. 3 Write a one-line function that takes a positive integer n and computes the sum

$$S(n) = \sum_{k=1}^{n} \frac{1}{k}.$$

Ex. 4 Write a MATLAB function that, given a positive integer n, checks whether the sum $S(n)$ defined by

$$S(n) = \sum_{k=1}^{n} \frac{1}{k(k+2)}$$

satisfies $S(n) < 3/4$ for all positive integers less than or equal to n. The function should return a vector of exactly n boolean values as output. Check your code by calling the function with several values of n.

Ex. 5 Create an anonymous MATLAB function that computes the value of $f(x)$, defined piecewise as:

$$f(x) = \begin{cases} x+1, & x \leq 0 \\ \sin(x), & x > 0 \end{cases}.$$

Ex. 6 Write a MATLAB anonymous function that can be used to compute an approximation to e^x using the first five terms of the Taylor series for the exponential around the point $x = 0$, that is:

$$e^x \approx f(x) = 1 + x + \frac{x^2}{2!} + \frac{x^3}{3!} + \frac{x^4}{4!}.$$

For a given point x, define the error in the approximation by calculating the absolute value of the difference $|f(x) - e^x|$. Plot this error versus x for all the points in the MATLAB vector x=0:0.2:2.

Ex. 7 The Taylor series for the function $\sin(x)$ around the point $x = 0$ is slightly more challenging. Similar to the previous exercise, write a MATLAB function that computes the value of $\sin(x)$ using seven terms in the series. Again, plot the error produced by this approximation for the points in the vector x=0:0.1:0.5. Remember that x here is in radians.

Ex. 8 Use the MATLAB help system to learn about the function `isprime`, then use this function together with `find` and the colon operator to produce a list of all the prime numbers less than or equal to $M = 30$.

Ex. 9 Use the MATLAB help system to learn about the function `mod`. Use this function to write a related function `isOdd` similar to the built-in function `isprime`. The function `isOdd` takes an integer argument and returns the value 1 if the argument is an odd number, otherwise it returns 0. Again, use `isOdd` together with `find` and the colon operator to produce a list of all the odd numbers less than or equal to $M = 21$.

Ex. 10 To investigate the roots of an equation of the form $f(x) = 0$, it may be helpful to plot the function $f(x)$ on an interval of interest. Use this method to find out if the following three functions have any root in the interval $[0, 2]$: $f_1(x) = e^x + x^2 - 5$, $f_2(x) = x^3 - 5x^2 + 2$ and $f_3(x) = \ln(x^2 + 2) - x^2$.

Ex. 11 Use the `diag` function provided in MATLAB to create a diagonal matrix with the same diagonal entries as the three-by-three matrix

$$A = \begin{pmatrix} 2 & 3 & 5 \\ -1 & 5 & 2 \\ 7 & 2 & 3 \end{pmatrix}.$$

Investigate the use of `triu` and `tril` to extract the upper- and lower-diagonal parts of A, respectively.

Ex. 12 Write MATLAB code that can be used to evaluate the product of the first line in a three-by-three matrix A with a column vector **x**. Check it out using the matrix A in the previous exercise and the vector $\mathbf{x} = [2\ 1\ 2]^T$. Then extend this code to make it able to handle the product of any line in A with the same vector. You can do this in at least two ways: first by extracting a line in the matrix A using the colon notation and multiplying with the implicit MATLAB matrix multiplication and second by using a `for` loop. Do it both ways and make sure you get the same result.

Ex. 13 Create a function that takes three vectors, p_1, p_2 and p_3, each with two components, as input. The vectors contain pairs of of coordinates $p_i = [x_i, y_i], i = 1, 2, 3$, for points in the plane. The function then creates a plot of the triangle with vertices at the three points.

Ex. 14 The *logistic equation* is a differential equation used to model populations that asymptotically approach a certain level K, known as the *saturation level*, as time evolves. Write a MATLAB anonymous function (you can call it pLogistic for example) of four variables that can compute the solution of the logistic equation, given by

$$p(t) = \frac{p_0 K}{p_0 + (K - p_0)e^{-rt}},$$

for any given set of values $\{p_0, K, r, t\}$. Make sure to write the function in such a way as to accept a vector of points for t; this variable represents time. Generate a set of one hundred points (tpts) equispaced in $[0, 5]$; using this set of points, plot the function pLogistic for the values $p_0 = 1$, $r = 0.4$, $K = 6$. Then change the value of p_0 to $p_0 = 3.2$ and plot the new function using the same values for the other variables. The two plots represent the evolution of two different populations that start at different levels at time $t = 0$.

Ex. 15 Using the timing tools presented in this chapter, verify the theoretical $\mathcal{O}(n^3)$ estimate for the order-n matrix multiplication operation count. To do so, use the **rand** MATLAB function to create two random matrices of order n. Multiply these matrices using the MATLAB built-in matrix multiplication operation and plot the resulting times versus n. Start with $n = 10$ and increase n by a factor of ten for each successive run; on most contemporary machines you should be able to increase it up to at least $n = 10,000$. Remember to suppress all output except for the timing commands. The estimate is valid as $n \to \infty$, so you won't obtain exactly that behavior, but you should be able to at least identify the trend. Please be aware that, for this experiment to succeed, it is best if MATLAB runs locally on your machine.

Ex. 16 The following numbers have a representation that uses only a finite number of digits in base two: 0.5, 0.75, 1.25, 5.775. Using the methodology presented in section 3.9, calculate their corresponding base-two representation. Verify your results by reverting the quantities you obtained back to base ten.

Ex. 17 Suppose that, together with computers using bits, we also have access to computers using *trits*. That is, the latter would use switches with three possible states, which can be denoted by 0, 1 and 2, to represent numbers. Stated another way, numbers are represented in base three on the second computer. Which of the following numbers, written in common decimal notation, would have a representation that uses only a finite number of switches on either one of these computers: 33, 221, 0.2, 0.3, $0.\overline{3} = 0.333\ldots$, 0.77? Write down the corresponding form for each of these numbers in the two cases.

Chapter 4

Solving Nonlinear Equations

One of the first tasks that can be tackled with just the basic tools introduced in the previous chapter is the solution of nonlinear equations in one scalar variable. This chapter discusses methods that have been historically developed for this purpose and are still in use today in their original or some slightly modified form. They are all iterative methods that start with a rough estimate of the root and refine this estimate until some degree of certainty that the estimate is sufficiently close to the root is achieved.

4.1 The Bisection Method for Root-Finding

Consider the problem of finding one **approximate root** of the trigonometric equation

$$f(x) = \sin(x) - 1.3\cos(1.45x) = 0.$$

For such nonlinear equations, a closed-form method of solution (i.e. a formula that provides the value of x satisfying the equation, such as the quadratic formula) is most of the time not available. In addition, apart from the ideal case of a unique solution, there may be more than one solution to such equations (as is the case here) or no solution at all. There simply is no general theory that can sort out the different cases. Therefore, it is always imperative that we obtain as much information about the equation we are trying to solve as possible. A good way to start here is by first plotting the function $f(x)$, say between $x = -10$ and $x = 10$:

> **MATLAB Script 4.1: Plotting the function**
> ```
> clear all              % clears all variables
> close all              % closes all figures
> x = -10:0.1:10;        % defines plot range
> y = sin(x) - 1.3*cos(1.45*x) ;   % creates f(x) values
> plot (x, y)            % plots data in x and y
> ```

This produces the graph shown in figure 4.1. Note that there are several roots in the interval under consideration, one of them being located in between

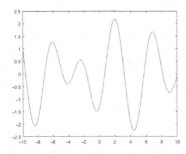

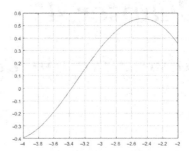

Figure 4.1: Function plot between $x = -10$ and $x = 10$.

Figure 4.2: Magnified plot of the region of interest.

$x = -4$ and $x = -2$. In order to zoom in on this particular root, one could use the following:

MATLAB Script 4.2: Zooming in on the region of interest

```
x = -4:0.05:-2;
y = sin(x) - 1.3*cos(1.45*x) ;
plot (x, y)
grid on      % creates grid in plot area
```

which produces figure 4.2. The added grid lines help us better locate the root.

To find any particular root of the equation, which will be denoted by x^*, one can start with an interval x_L, x_R that encloses that particular root and will be referred to as a *bracket*, as shown in figure 4.3.

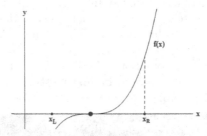

Figure 4.3: Graph of a function f(x) and bracket points x_L and x_R.

Under some mild assumptions (left for the reader to figure out) on the function $f(x)$, one may be certain that there's at least one zero in (x_L, x_R) if $f(x_L) * f(x_R) < 0$. To zoom in on the root, a valid strategy is to start by locating the mid-point of the interval, x_M say, and construct the next bracket around the solution as either

(a) $[x_L, x_M]$ if $f(x_L)f(x_M) < 0$, or

(b) $[x_M, x_R]$ if $f(x_M)f(x_R) < 0$.

Upon doing this until, hopefully, some level of confidence that we are close to the root is achieved, the midpoint of our last (smallest) bracket will be a reasonable approximation to the true zero of the function. This, in a nutshell, is the *bisection method* for root-finding.

4.2 Convergence Criteria and Efficiency

The question that arises immediately is: how can one be confident enough that a good approximation for the root has been obtained? This is always an important question for iterative methods like bisection. To answer the question, it is useful to remember that at the actual root $x = x^*$ one knows that $f(x = x^*) = 0$. For other values of x, around x^*, the value of $f(x)$ will not be exactly zero, but should be very small in absolute value and should get smaller as we approach x^*. Therefore $|f(x)|$ is a readily-available measure for the accuracy of our computation. Since this value should be zero when the iterative process converges, it is usually known as the *residual* at the current iteration, or simply the residual. Usually one wants to drive the residual below a certain, appropriately chosen, accuracy threshold which will be denoted here by α. Thus, from this perspective, the iteration can be stopped when $|f(x)| < \alpha$. For iterative methods it is also appropriate to, most of the time, also impose a maximum limit on the number of iterations (bracket halving) performed.

At this point, let us turn our attention to creating MATLAB code that implements the ideas from the previous discussion in order to produce an approximation to the root. Such code will contain a number of commands in succession, making it rather inconvenient to be typed at the command prompt. Instead, one can write these commands in a file known as a MATLAB *script*. To do so, use the MATLAB environment menu ("New/Script") to create a new script file which will save the code implementing the bisection method. Doing so will open the MATLAB editor, which appropriately formats and highlights the commands; the file can then be saved, for example under the name `bisection1.m`. According to our discussion, a first shot at it with $\alpha = 10^{-5}$ might look as shown in the following script:

MATLAB Script 4.3: Bisection Method Script, version 1

```
clear;          % clears all variables
xL = -4 ;       % initial left boundary
xR = -2 ;       % initial right boundary

for j = 1:10
    xM = (xL + xR)/2 ;
    fM = sin(xM) - 1.3*cos(1.45*xM) ;
    fL = sin(xL) - 1.3*cos(1.45*xL) ;
    fR = sin(xR) - 1.3*cos(1.45*xR) ;
    if fM*fL < 0
        xR = xM;
    elseif fM*fR <0
        xL = xM;
    end
    if abs(fM) < 10^(-5) % checking accuracy
        break     % this exits the (innermost) loop!
    end
end
```

Upon saving the file, the script can be run by either clicking the "Run" button or typing its name (without the extension) at the command prompt as shown below:

>> bisection1

For either variant to work, the file must be in MATLAB's active path. Notice that, when a script file runs, the variables it defines become part of the workspace.

While the above script 4.3 does implement the ideas discussed so far, there are several improvements that one can think of. Consider, in particular, the fact that this piece of code must produce a result within a reasonable amount of time; issues related to efficiency immediately arise once one moves away from the idealized mathematical-problem setting. First of all however, notice that the expression for the function was typed three times. This is rather a matter of poor style: one can define an anonymous function and use that function instead. Second, let us suppose that the initial interval has been carefully selected so that we are certain that one root, and one root only, resides therein. Here we accomplished this by zooming into the graph of the function. Then if the first test that the code makes fails, i.e. if $f(x_M) * f(x_L) \geq 0$, one can be certain that the root will be in the other half of the interval. Therefore, the other test $f(x_M) * f(x_R) < 0$ is not needed and thus neither is there any reason to compute $f(x_R)$. With this in mind, the included loop may be more efficiently written as:

MATLAB Script 4.4: Bisection Method Script, version 2

```
clear all;      % clears all variables
xL = -4 ;     % initial left boundary
xR = -2 ;     % initial right boundary
f = @(x) sin(x) - 1.3*cos(1.45*x) ;

for j = 1:10
    xM = (xL + xR)/2 ;
    fM = f(xM) ;
    fL = f(xL) ;
    if fM*fL < 0
        xR = xM;
    else
        xL = xM;
    end
    if abs(fM) < 10^(-5)
        break
    end
end
```

The reader is invited to think about this change: is it totally error prone? Are there any scenarios that have been neglected but should be taken into account? On the other hand, are there other changes that may make your program perform even less computations? This is an important skill to develop for scientific computing.

Apart from the residual accuracy criterion, the bisection method offers another way to check convergence, since it provides an upper bound on the distance between the estimate for the root (the midpoint of the current bracket) and the actual root. Let us denote by x_M^1 the first midpoint, that is the midpoint of the original bracket where the root has been located. Obviously one has:

$$|x_M^1 - x^*| \leq \frac{x_R - x_L}{2},$$

where equality was allowed, in case one of the endpoints of the bracket happens to be the root. Denoting by x_M^k the midpoint after k subdivisions, it should be clear that

$$|x_M^k - x^*| \leq \frac{x_R - x_L}{2^k}. \tag{4.1}$$

Suppose now that the goal is to have the root within a certain distance, to be called herein tolerance and denoted by T, from the current estimate. Then asking that

$$\frac{x_R - x_L}{2^k} \leq T$$

hold will ensure that $|x_M^k - x^*| \leq T$. Thus one can solve for the number of subdivisions required by the tolerance condition once the initial bracket $[x_L, x_R]$ and the value of T are known:

$$k \geq \log_2\left(\frac{x_R - x_L}{T}\right).$$

Both convergence criteria can be incorporated in the code in a number of ways. Below is a variant that iterates a maximum of maxIter times but ends the iterative process once both convergence criteria are met. Alternately, one can compute the number of iterations that are needed to reach the prescribed tolerance T before the loop starts and ask that the loop runs at least that many iterations.

MATLAB Script 4.5: Bisection Method Script, version 3

```
clear;        % clears all variables
xL = -4 ;     % initial left boundary
xR = -2 ;     % initial right boundary
f = @(x) sin(x) - 1.3*cos(1.45*x) ;
maxIter = 1000;
T = 10^(-6);
for j = 1:maxIter
    xM = (xL + xR)/2 ;
    fM = f(xM) ;
    fL = f(xL) ;
    if fM*fL < 0
        xR = xM;
    else
        xL = xM;
    end
    if abs(fM) < 10^(-5) & (xR - xL) < T
        break;
    end
end
xM  % xM = -3.3714
fM  % fM =  0.00000072176
j   % j = 21
```

It is instructive to also look at a straightforward while loop version of this code. Checking for convergence needs to be done at the start of the loop, hence the involved quantities must be readily available. Therefore, in order to avoid repeated computation of the initial xM and fM values, it is necessary to rearrange the calculations inside the loop. Also note the way the condition changes from a conjunction to a disjunction and the fact that the while loop doesn't automatically keep track of the number of iterations. To adjust for the latter problem one may define a counter that needs to be incremented at each pass through the loop.

MATLAB Script 4.6: Bisection Method Script, version 4

```
clear;          % clears all variables
xL = -4 ;       % initial left boundary
xR = -2 ;       % initial right boundary
f = @(x) sin(x) - 1.3*cos(1.45*x) ;
T = 10^(-6);
xM = (xL + xR)/2 ;
fM = f(xM) ;
while abs(fM) > 10^(-5) || (xR - xL) > T
    fL = f(xL) ;
    if fM*fL < 0
        xR = xM;
    else
        xL = xM;
    end
    xM = (xL + xR)/2 ;
    fM = f(xM) ;
end
xM
fM
```

4.3   Scripts and Function Files

As seen above, the codes we write can be saved as **script files** which can then
be run repeatedly. Here are some of the important rules governing the use of
scripts:

1. Script files must have valid variable names.

2. They must be accessible in the MATLAB path or the current working
   directory. The working directory is displayed by MATLAB under the
   menu items, and can be changed by clicking on it and selecting a different
   directory from the drop-down list.

3. Variables with the same names as scripts should not be used. Such variables would have precedence over scripts, so MATLAB will not see a
   script as long as a variable with the same name exists.

4. Assigning scripts to variables or passing arguments to scripts (as you
   would to a function) is not allowed;

   ```
   >> x = bisection1;
   >> bisection1(3)
   ```

   are both bad commands.

5. Script files will overwrite values of your variables in the environment if
   they use variables with the same names. This may be desirable, but it

may also lead to problems if done unintentionally. To prevent this from happening, one may start their scripts with the command `clear all`. This will remove all variables from the environment, so script variables don't inadvertently inherit variables previously defined in the command window.

6. Scripts are particularly useful as a high-level command center which then delegates the work to the other units of code. As will be seen, these latter units of code will usually be functions.

MATLAB functions can be defined in two ways: as **anonymous functions** or as **file functions**. Anonymous functions have been already encountered in Chapter 3. The concept of a function can be generalized to allow us to obtain more than the root within one particular bracket with our bisection code. More precisely, suppose that one wants to run the same bisection code but with different initial brackets. In our setup so far, in order to do so, one needs to modify the code in the script file. This is not the best strategy; after some time, one will probably forget at least some details about their code and may therefore introduce errors. MATLAB file functions allow us to pre-package this code in a sort of black box that can still accept different initial brackets as inputs. Start by creating a function file, `bisection.m`, by accessing ("New/Function") in the menu. If using the `while` loop version, for example, the contents of the file would look as below:

MATLAB Script 4.7: Bisection Method Function, version 1

```
function [ result ] = bisection(xL, xR)
f = @(x) sin(x) - 1.3*cos(1.45*x) ;
T = 10^(-6);
xM = (xL + xR)/2 ;
fM = f(xM) ;
while abs(fM) > 10^(-5) || (xR - xL) > T
  fL = f(xL) ;
  if fM*fL < 0
    xR = xM;
  else
    xL = xM;
  end
  xM = (xL + xR)/2 ;
  fM = f(xM) ;
end
result = xM
end
```

Once this is saved in the working directory, one can use it with different brackets, for example by typing at the command line:

```
>> root1 = bisection(0.1, 1.5)
root1 = 0.71757
```

```
>> root2 = bisection(2.0, 4.0)
root2 =   3.2124
```

Even more interestingly, one may want to pass as argument a different function, other than $f(x)$, to the bisection.m code. In that case, the code will perform bisection on this actual function supplied in the input argument list. In other words, MATLAB allows functions as input arguments that can be supplied to other functions. To do that in the most proper way, let us first note that there are two ways to evaluate a function in MATLAB:

```
>> f = @(x) sin(x) - 1.3*cos(1.45*x);
>> f(0.1)                        % infix notation
>> feval(f, 0.1)                 % functional notation
```

The functional notation, using feval, has the advantage to clearly distinguish between functions and vectors. That is, when using the infix notation, there is at times no clear way to tell apart a function evaluation from accessing a vector element:

```
>> v(2)                 % is v a function or a vector?
```

A MATLAB function that is handed another function as an input argument does not always have a way[1] to determine whether its incoming input is actually a function or a vector. It is therefore good practice to evaluate the latter function using feval inside the body of the first function. The latter script can therefore be modified to read as below:

MATLAB Script 4.8: Bisection Method Function, version 2

```
function [ result ] = bisection(f, xL, xR)
T = 10^(-6);
xM = (xL + xR)/2 ;
fM = f(xM) ;
while abs(fM) > 10^(-5) || (xR - xL) > T
  fL = feval(f, xL) ;
  if fM*fL < 0
    xR = xM;
  else
    xL = xM;
  end
  xM = (xL + xR)/2 ;
  fM = feval(f, xM) ;
end
result = xM
end
```

Hence to use it one could use a script that contains, for example, commands like the following:

---

[1] This is a consequence of MATLAB passing arguments only by value, instead of also passing by reference as in other languages.

```
>> f1 = @(x) sin(x) - 1.3*cos(1.45*x);
>> xL1 = 2.0;
>> xR1 = 4.0;
>> root1 = bisection(f1, xL1, xR1)
>> f2 = @(x) x^2 - 3*x + 2; %different function
>> xL2 = 1.1;
>> xR1 = 3.2;
>> root1 = bisection(f2, xL2, xR2)
```

In the above code snippet, both functions passed as arguments to bisection are anonymous functions. If the function to be used as an argument is instead defined in a function file, MATLAB requires us to include the name of the file in between quotes when using it as an argument. Such an use would look as in the example below:

```
>> root3 = bisection('myFunctionFileName', ...
            xL, xR)
```

This may be a good place to mention a peculiarity of MATLAB functions[2], which distinguishes them from their stricter mathematical counterparts. MATLAB functions are allowed to produce multiple outputs. The outputs are listed in between square brackets following the name of the function. By default, only the first output is returned when the function name is invoked; to obtain all outputs, one needs to use a vector that receives the desired quantities. The example code below shows a classical example of such a function,

MATLAB Script 4.9: Two Means Function

```
function[amean, gmean] = bothMeans(a,b)

amean = (a+b)./2;    % arithmetic mean
gmean = sqrt(a.*b);  % geometric mean

end
```

which can then be called in the following ways:

```
>> [amean, gmean] = bothMeans(1,4)
amean = 2.5
 gmean = 2
>> bothMeans(1,4)
ans = 2.5 %only first output returned
```

Allowing for more than one output is a tremendous advantage: as will be seen throughout this text, it is many times just natural to use the same input

---

[2]Code modules that returned several outputs were initially known as subroutines. They became known as functions after the introduction of the very successful C language.

arguments to calculate more than just one quantity of interest. Storing and returning the results of such calculations as function outputs can save time and make our code better structured and easier to read and maintain.

4.4   The False Position Method

This method, also known as *regula falsi*, shares the simplicity of the bisection method. In most cases it may converge slightly faster than bisection, however, because it attempts to account for different rates of change of the function within the current bracket. Specifically, instead of using the midpoint of the current bracket in order to divide it into the next two possible brackets, the method chooses the point that is obtained by using a secant to the graph of the function. This amounts to approximating the function by a straight line, then finding the intersection of this line with the horizontal axis instead.

Suppose, for concreteness, that the current bracket is $[x_L, x_R]$. The points $P_L = (x_L, f(x_L))$ and $P_R = (x_R, f(x_R))$ determine a straight line, which will intersect the horizontal axis at some point $\tilde{x}_M$ given by:

$$\tilde{x}_M = \frac{x_L f(x_R) - x_R f(x_L)}{f(x_R) - f(x_L)}.$$

Upon calculation of $\tilde{x}_M$, a test similar to the one used in the bisection method selects either the left side of the initial bracket, $[x_L, \tilde{x}_M]$, or the right side $[\tilde{x}_M, x_R]$. The selected sub-interval becomes the next bracket, and the iterative process is repeated until convergence.

4.5   The Newton-Raphson Method for Root-Finding

This more powerful method uses additional information to find the root of a given function. It does require the function to be differentiable and is usually much faster than the bisection method, but it may also fail in some cases. Instead of an initial bracket, one starts with just one initial guess point, x_0. The basic idea is to approximate $f(x)$ by its tangent at this point, a process known as linearization. The intersection of this tangent with the x-axis becomes the next approximation to the root, as shown in figure 4.4. The process may either continue by having this approximation take the place of the initial guess, or stop if the residual dropped below an acceptable threshold.

The slope of a function at a generic point x is given by the derivative $f'(x)$. Therefore, a straight line passing through $(x_0, f(x_0))$ with slope $f'(x_0)$ has the equation

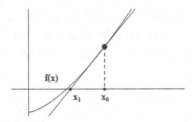

Figure 4.4: A function $f(x)$ with its tangent line at x_0.

$$\frac{y(x) - f(x_0)}{x - x_0} = f'(x_0),$$

which is simply a statement of the fact that the rise divided by the run is equal to the slope. Thus, the goal is to find x_1, which is the point where $y(x_1) = 0$. This leads to:

$$0 = y(x_1) = f'(x_0)(x_1 - x_0) + f(x_0),$$

so that, by rearranging, one obtains:

$$-f(x_0) = f'(x_0)(x_1 - x_0),$$

and therefore,

$$x_1 = x_0 - \frac{f(x_0)}{f'(x_0)}. \tag{4.2}$$

As an iterative process, the goal is to produce a sequence of approximations $x_0, x_1, \ldots, x_n, \ldots$, where for each $n = 0, 1, \ldots$, the next x_{n+1} point is obtained by thinking of x_n as the initial point and using the above relationship. This can then be expressed by the following equation for the move from x_n to x_{n+1}:

$$x_{n+1} = x_n - \frac{f(x_n)}{f'(x_n)}, \quad n = 0, 1, \ldots \tag{4.3}$$

Here is a straightforward initial implementation in a script file, which supposes that the functions f and fprime, together with the initial guess x(1), are defined before the start of the loop:

MATLAB Script 4.10: Newton-Raphson Method

```
clear;
f = @(x) ...; %the function
fprime = @(x) ...; %the derivative
x(1) = ...;          % initial guess
for i = 1:10
    x(i+1) = x(i) - f(x(i))/fprime(x(i));
    fval = f(x(i+1));
    xval = x(i+1);
    if abs(fval) < 10^(-5)
        break
    end
end
xval                % print value of root
fval                % print function value
```

Note that usually there is no real need to keep all the elements of the vector x; one may just keep the two most current iterations. Also note that no precautions have been taken for the case when the value of the derivative becomes of the order of the machine precision or lower. This may lead to division-by-zero errors and should be prevented. These improvements, together with extra flexibility in the check for convergence and maximum number of iterations, are left for the reader.

4.5.1 Convergence of the Newton-Raphson Method

The Newton-Raphson method usually converges very fast if a good initial guess is used; it exhibits what is known as *quadratic convergence*. To understand the ideas involved, let us look at the Taylor series development of $f(x)$ around x_n. If keeping only the linear term, the expansion has the form:

$$f(x) = f(x_n) + f'(x_n)(x - x_n) + \frac{f''(\xi)}{2}(x - x_n)^2$$

where the remainder is computed at some point ξ in between x and x_n. Since we are searching for the root, let us set $x = x^*$ so the value of $f(x^*)$ on the left-hand side is zero. With this in mind, as long as $f'(x_n)$ is not zero, one can divide by this value and move some terms to the left-hand side to end up with the following expression:

$$x^* - \left(x_n - \frac{f(x_n)}{f'(x_n)}\right) = -\frac{f''(\xi)}{2f'(x_n)}(x^* - x_n)^2$$

which now produces, taking 4.3 into account:

$$x^* - x_{n+1} = -\frac{f''(\xi)}{2f'(x_n)}(x^* - x_n)^2.$$

Using the absolute value on both sides of this equation leads to:

$$|x^* - x_{n+1}| \leq C|x^* - x_n|^2, \ C > 0 \tag{4.4}$$

where the positive constant C is the upper bound of the fraction $|f''(\xi)/2f'(x_n)|$ for the interval under consideration. Notice that introducing this constant makes the implicit assumption that the derivative doesn't take the value zero within this interval. The relationship above tells us that the distance between the actual root and its approximation at step $n+1$ is, at most, proportional to the square of the distance in the preceding step n, hence the term quadratic convergence. It should be contrasted with the corresponding relationship for the bisection method, for which equation (4.1) implies

$$|x^* - x_M^{n+1}| \leq \frac{1}{2}|x^* - x_M^n|. \tag{4.5}$$

This shows that the bisection method exhibits what is known as *linear convergence*.

Figure 4.5 shows how the error decreases with each iteration for Newton's method, as compared to the bisection method. The test is done on the function $f(x) = x^2 - 5$. In order to start with the same residual, the bisection method uses an initial bracket $[x_L = 0, x_R = 10]$ while Newton's method starts with the initial guess $x_0 = 5$. The graph uses a logarithmic scale on the y-axis, obtained using the `semilogy` command, to show more detail for small residual values. This is common practice for such a *convergence plot*.

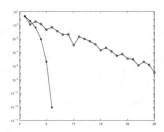

Figure 4.5: Residual convergence for Newton's method versus bisection.

4.5.2 Recursive functions and optional arguments

Since this chapter introduced MATLAB functions, it offers good grounds to look at other possibilities that can sometimes simplify the translation our ideas into code. Consider, as a first example, the task of writing a function that computes and returns as its single output the value of the factorial   .

$$n! = 1 \cdot 2 \cdot 3 \ldots (n-1) \cdot n$$

A simple, straightforward implementation might look as the piece of code below:

Matlab Function 4.1: Non-Recursive Factorial Function

```
function [ val ] = fact1(n)
val = 1;

for i = 1:n
    val = val * i;
end

end
```

Note that this function will work for the input $n = 0$, in which case it returns the value one, since the body of the loop will not be executed. In fact, the same is true if this function is called with a negative integer as argument.

One can also think about the factorial in a different way, starting from the other end of the terms involved in the product. That is, one could obtain the factorial as:

$$n! = n \cdot (n - 1)!$$

This is obviously very handy in case someone already knows the value of $(n - 1)!$ since they can get $n!$ for potentially large n with only one multiplication. This remark leads us to consider a function that computes $n!$, in a somewhat lazy fashion, by first delegating the task of computing $(n - 1)!$ then just using that value and computing the single extra product needed. Obviously, one couldn't be asking for $(n - 1)!$ forever; as n becomes lower, say $n = 1$, then $1! = 1$ and we can stop here by not delegating any more and instead defining:

```
(n-1)! = 0! = 1
```

This idea leads to another version of the factorial function, one which includes a call to itself:

Matlab Function 4.2: Recursive Factorial Function

```
function [ val ] = fact2(n)
val = 1;

if (n == 0)
    val = 1;
else
    val = n * fact2(n-1); % recursive call
end

end
```

This `fact2` version of the factorial, which computes the factorial by the use of an evaluation of the same function `fact2` but with a different argument, is known as a *recursive function* implementation. All such recursive functions need to handle a base case (the case $n = 0$ in this example) where the recursive call does not appear any more; otherwise the process would go on indefinitely.

Recursive functions are not very common in scientific computing because of the high computational costs associated with them (in particular, dynamic memory allocation costs). However, they may make sense in some instances where recursive algorithms are natural. Shown below is a possible recursive implementation of Newton's method which also aims at returning the list of all the successive approximations to the root that are computed along the way to convergence.

Matlab Function 4.3: Recursive Newton's Method

```
function [xStar, xL] = newt_RecV1(f, fp, xN, xList)
% Recursive implementation of Newton's Method
if abs( feval(f, xN) ) < 10^(-5)
    xStar = xN;
    xL = xList;
    return;
else
    xNp1 = xN - feval(f, xN) / feval(fp,xN);
    %xStar = xNp1;
    xList = [ xList xNp1 ];
    xL = xList;
    [xStar, xL] = newt_RecV1(f, fp, xNp1, xList);
    % newt_Rec(f, fp, xNp1, xList);
end
end
```

This implementation has a slight problem, in that the initial guess needs to be used both as an argument by itself and also to initially populate the list of approximations. While this can be avoided in various ways, one way to avoid it is by using an optional argument as shown in MATLAB Function 4.4. A script that calls both is also included to show the difference in their use.

Matlab Function 4.4: Recursive Newton's method with optional argument

```
function [xStar, xL] = newt_RecV2(f, fp, xN, xList)
% Recursive Newton's method with optional argument
if ~exist('xList', 'var')
    xList = [xN];
end
if abs( feval(f, xN) ) < 10^(-5)
    xStar = xN;
    xL = xList;
    return;
else
    xNp1 = xN - feval(f, xN) / feval(fp,xN);
    %xStar = xNp1;
    xList = [ xList xNp1 ];
    xL = xList;
    [xStar, xL] = newt_RecV2(f, fp, xNp1, xList);
end
end
```

Matlab Function 4.5: Recursive Newton Script

```
clear all;
format long;
f = @(x) x.^2 - 7;
fp = @(x) 2*x;
% Note redundancy in initial guess use below
[root, xList] = newt_RecV1(f, fp, 1, [1] )
% This first call of V2 only needs three arguments
[root, xList] = newt_RecV2(f, fp, 1 );
root
xList
```

4.6   Fixed Point Iteration

Iterative methods are ubiquitous in scientific computing problems arising in science and engineering. A quick glance at the concept of fixed point iteration, undertaken in this section, offers a very useful way to understand their convergence. Our discussion will center around Newton's method, which can be framed as a fixed-point iteration process. The basic form of a fixed-point iteration scheme is:

$$x_{k+1} = g(x_k),$$

where the *iteration function* g constitutes the update rule to be applied. For Newton's method the function g is related to, but clearly not the same as, the function f; more exactly, it is given by

$$g(x) = x - \frac{f(x)}{f'(x)}.$$

Starting with an initial value x_0, an iterative process constructs a sequence:

$$x_1 = g(x_0)$$
$$x_2 = g(x_1)$$
$$\vdots$$

A *fixed point* for the iterative scheme $x_{k+1} = g(x_k)$ is a point x^* for which $x^* = g(x^*)$. Stated otherwise, an iteration that reaches a fixed point does not advance any longer. A graphical approach to understanding the behavior of the scheme starts by considering the following two functions:

$$f_1(x) = x \qquad f_2(x) = g(x).$$

Using these functions, it can be seen that fixed points occur if and only if $f_1(x^*) = f_2(x^*)$, which is a different way to state that $x^* = g(x^*)$. That is, the fixed point resides at the intersection of the graphs of the two functions. The most common situations that occur during fixed-point iteration can then be viewed geometrically. Figures 4.6 through 4.9 show four of these scenarios.

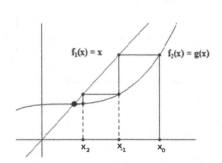

Figure 4.6: Monotonic convergence.   **Figure 4.7:** Monotonic divergence.

As these figures suggest, if a choice is possible, the iteration function $g(x)$ needs to be selected with care in order for the iterative process to converge in the fastest manner. For an example, consider the simple task of finding the square root of two. One function that has a fixed point at $x^* = \sqrt{2}$ is $g_1(x) = 2/x$. An attempt to obtain x^* using this function in a fixed-point iteration scheme fails, irrespective of the starting point, due to oscillatory behavior. A simple calculation with $x_0 = 1$ highlights the problem: the sequence of iterates is $\{1, 2, 1, 2, \ldots\}$. This behavior is shown in figure 4.10 for the starting point $x_0 = 1.6$ and is to be contrasted with the behavior that ensues when the same

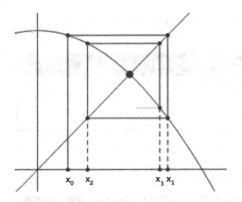

Figure 4.8: Oscillatory convergence.   **Figure 4.9:** Oscillatory divergence.

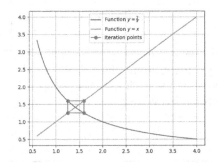
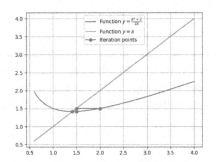

Figure 4.10: Oscillatory behavior for the iteration function $g_1(x) = 2/x$.

Figure 4.11: Convergence for the iteration function $g_2(x) = (x^2+2)/(2x)$.

square root calculation is attempted using Newton's method on the function $f(x) = x^2 - 2$. In this latter case it can be easily verified that the corresponding iteration function is $g_2(x) = (x^2+2)/(2x)$ and the iteration converges, as seen in figure 4.11 for $x_0 = 2$.

4.7   MATLAB® Built-in Functions

To solve nonlinear equations, MATLAB offers the built-in function `fzero`. It uses a combination of the bisection, secant and an inverse interpolation method. The user does not need to specify an initial bracket, but may do so for improved performance. The script below uses this function to find the zero of $f(x) = x^2 - 7$.

MATLAB Script 4.11: Using fzero with an initial guess

```
f = @(x) x^2 - 7;
x0 = 1.0;
x = fzero(f,x0) % output: 2.6458
```

Different options may be used in order to ask `fzero` to behave differently. For example, in order to use an initial bracket and print an account of how the iterations proceed, one can use the following code:

MATLAB Script 4.12: Using fzero with an initial bracket

```
f = @(x) x^2 - 7;
x0 = [1.0, 7.0];
options = optimset('Display','iter'); % print iterations
x = fzero(f, x0, options)
```

An initial bracket may speed up the computation considerably. In this simple example, the function is evaluated thirty four times when script 4.11 is executed, but only eleven times when using script 4.12.

4.8   Exercises

Ex. 1 Find the first two positive roots of the function

$$f(x) = \sin(x + 0.5e^{-x}) + \cos(0.4x).$$

Use the bisection method with your own choice of initial brackets.

Ex. 2 Use Newton's method to solve the equation

$$x + \sin(x) = \cos(x).$$

Use the following two initial guess points: (a) $x_0 = 0.2$; (b) $x_0 = 0.8$. In both cases, converge the residual to an accuracy $\alpha = 10^{-5}$, then plot the residual versus the iteration number using a semi-logarithmic scale for the vertical axis.

Ex. 3 Consider applying the bisection method to the problem above, using an intial bracket $[x_L, x_R] = [2, 4]$. How many iterations are required to obtain the root located in this interval with a tolerance $T = 10^{-7}$?

Ex. 4 Write a MATLAB code for a function that implements the false position method. Similar to the code in 4.8, the function should take as arguments the two points defining the initial bracket and the function for which one needs to find the root. Solve the problem above with both this and the bisection method and compare the results.

Ex. 5 Use the built-in MATLAB function `fzero` to obtain the solution to

$$f(x) = e^x - \sin(x) - 4x = 0$$

Call the function in two ways, both with and without specifying the initial bracket $[1, 3]$.

Ex. 6 Think about Newton's method and its disadvantage: it requires knowledge of the derivative of the function. Develop a method that circumvents this problem by using an approximation to the derivative obtained using a secant to the graph of the function. That is, suppose that you start with two points x_0 and x_1. Your first approximation to the root, x_2, will be the intersection of the secant through $(x_1, f(x_1))$ and $(x_0, f(x_0))$ with the x-axis. Find the equation of this line, express x_2 as a function of x_0 and x_1, then generalize from $\{x_0, x_1, x_2\}$ to $\{x_{k-1}, x_k, x_{k+1}\}$.

Ex. 7 Implement the method described in **Ex. 6** in MATLAB. Use it to evaluate $\sqrt{18}$. Start with $x_0 = 0$ and $x_1 = 5$.

Ex. 8 As discussed in this chapter, when setting up fixed-point iteration one has to be aware that, while there may exist multiple functions that have the same fixed point, fixed-point iteration will not always converge with all these functions. Study the convergence of the following functions: $g_1(x) = (x^2 + a)/(2x)$, $g_2(x) = a/x$ and $g_3(x) = 2x^3/(a + x^2)$ for finding the value of $\sqrt{a}$. Start with the value $x_0 = a$ and perform nine iterations for each function.

Ex. 9 Write the MATLAB code for a recursive function that computes the n-th member f_n of the Fibonacci sequence defined by $f_1 = 0$, $f_2 = 1$ and $f_n = f_{n-1} + f_{n-2}$ for $n = 3, 4, \ldots$.

Ex. 10 Write a MATLAB code for a recursive function that generates and prints all the permutations of a set of n elements. For convenience, you may label the elements with a corresponding natural number.

Chapter 5

Systems of Equations

The solution of systems of either linear or nonlinear equations is a problem that arises very often in computational science. This is due, among other reasons, to the fact that the systems of differential equations used to model a variety of complex natural phenomena ultimately lead to such systems as an expression of the relationships between the variables of interest. These systems can be very large, with millions of unknowns, therefore it is imperative that efficient methods be used for their solution.

5.1 Linear Systems

As will be seen later on in this chapter, solving nonlinear systems of equations can ultimately be reduced to solving linear systems. Two important pathways can be taken toward this latter end. One may use a direct solution method, which is guaranteed to produce the exact solution, whenever such solution exists, if all calculations can be done without any error. Alternately, one can use an iterative solution method, which usually produces approximate results within some accuracy threshold. In addition, when the calculations are subject to round-off errors, methods in these two categories are sometimes combined. For example, an iterative method may be used to refine the result, obtained by a direct method, that is subject to errors due to finite precision arithmetic. The best known direct method is Gauss elimination, which has been introduced in Chapter 2. The presentation below starts by first looking at a few extra aspects of this method that make it more manageable, then discussing one of its many MATLAB implementations.

5.1.1 Gauss Elimination Revisited

Consider a system of three equations in three unknowns, whose solution can be thought of as the point of intersection of three planes:

$$\begin{cases} x + y + z = 1, \\ 2x + 3y + 4z = 1, \\ 2x - y - 5z = 6. \end{cases}$$

DOI: 10.1201/9780429262876-5

Written in matrix form, this means that one is looking for the vector $\mathbf{x}$ such that $A\mathbf{x} = \mathbf{b}$, where the matrix A, known as the system matrix, the right-hand side vector $\mathbf{b}$ and the vector of unknowns, respectively, are:

$$A = \begin{bmatrix} 1 & 1 & 1 \\ 2 & 3 & 4 \\ 2 & -1 & -5 \end{bmatrix}, \qquad \mathbf{b} = \begin{bmatrix} 1 \\ 1 \\ 6 \end{bmatrix}, \qquad \mathbf{x} = \begin{bmatrix} x \\ y \\ z \end{bmatrix}.$$

One way to organize the ensuing calculations when performing Gauss elimination is by using the extended, or *augmented*, matrix $[A\,|\,\mathbf{b}]$ obtained by adding the right-hand side vector $\mathbf{b}$ as an extra column to the system matrix A. The calculation can then proceed with its two stages, as previously discussed. The first stage, *forward elimination*, attempts to produce zeros below the diagonal elements of the matrix A, which are known as the *pivots*. In the second, *backward substitution* stage, the unknowns are solved for, starting from the last one and proceeding toward the first. The forward elimination stage for this system can be represented schematically in the following succession of arrays that show the state of the system matrix and the right-hand side vector after each elimination:

$$\begin{bmatrix} \mathbf{1} & 1 & 1 & | & 1 \\ 2 & 3 & 4 & | & 1 \\ 2 & -1 & -5 & | & 6 \end{bmatrix} \rightarrow \begin{bmatrix} \mathbf{1} & 1 & 1 & | & 1 \\ 0 & \mathbf{1} & 2 & | & -1 \\ 0 & -3 & -7 & | & 4 \end{bmatrix} \rightarrow \begin{bmatrix} \mathbf{1} & 1 & 1 & | & 1 \\ 0 & \mathbf{1} & 2 & | & -1 \\ 0 & 0 & \mathbf{-1} & | & 1 \end{bmatrix}$$

In the elimination calculations pictured above, the pivots are shown in bold. After the final elimination step, the last equation obviously reads $(-1)z = 1$, so $z = -1$. Now the backward substitution stage starts; using this value in the second equation $y + 2z = -1$ produces $y = 1$, after which $x = 1$ follows from the first equation.

It should be obvious that the only operations allowed in the forward elimination process should be those that do not alter the information contained in the original equations that make up the system. These operations are known as *elementary row operations*. These are the addition (or subtraction) of one row from another, as well as multiplication and division of rows by non-zero scalars. As has been shown in Chapter 2, the forward Gauss elimination stage amounts to factoring the matrix A into the lower-triangular part L and upper-triangular part U such that:

$$A = LU = \begin{bmatrix} a_{11} & a_{12} & a_{13} \\ a_{21} & a_{22} & a_{23} \\ a_{31} & a_{32} & a_{33} \end{bmatrix} = \begin{bmatrix} 1 & 0 & 0 \\ m_{21} & 1 & 0 \\ m_{31} & m_{32} & 1 \end{bmatrix} \begin{bmatrix} u_{11} & u_{21} & u_{31} \\ 0 & u_{22} & u_{23} \\ 0 & 0 & u_{33} \end{bmatrix},$$

where the entries in the lower-triangular part L are the multipliers needed in the elimination process, while the upper-triangular matrix U is the matrix obtained from A after the forward elimination stage is complete. For the example above the two factors can be seen to be:

$$A = LU = \begin{bmatrix} 1 & 0 & 0 \\ 2 & 1 & 0 \\ 2 & -3 & 1 \end{bmatrix} \begin{bmatrix} 1 & 1 & 1 \\ 0 & 1 & 2 \\ 0 & 0 & -1 \end{bmatrix}.$$

To solve a single system in MATLAB, one uses a predefined infix operator:

```
>> A = [1 1 1; 2 3 4; 2 -1 -5];
>> b = [1; 1; 6];
>> x = A \ b; % solution
```

usually referred to as "A under b". Equivalently, since solving $Ax = b$ is equivalent to solving $LUx = L(Ux) = b$, the decomposition can be exploited when one needs to solve several systems with the same system matrix A but with different right-hand side vectors b. Supposing that the factors L and U are known, for example from the solution of a first system with a given matrix A, let $Ux = y$. The unknown vector x is not known at this stage, so this introduces a new, unknown, vector y. Nevertheless, it follows that we can solve for this new unknown easily. It is obtained as the solution of the system:

$$Ly = b,$$

which can be seen to be much easier than solving the original system, as it is similar to backward substitution. Once y is known, the initial unknown x can be solved for as usual, by backward substitution, from the triangular system of equations:

$$Ux = y,$$

where the vector y is now known, since it has been solved for in the previous stage.

Solving an upper- or lower-triangular system has an operation count $\mathcal{O}(n^2)$ versus the $\mathcal{O}(n^3)$ needed for a complete Gauss elimination. Thus, if one needs to solve multiple systems with the same matrix A but different right-hand side vectors b, it is a good idea to find and store the factors of A, L, and U. Obtaining these factors in MATLAB is straightforward with the lu function:

```
>> [L, U] = lu(A)    %computes L & U factors of A
```

Nevertheless, in this form the MATLAB command does not always directly return the two factors of A. It returns instead the factors of a related matrix that has the same rows as A, but possibly in a reshuffled position. This is because the lu command is more complex than the simple LU-decomposition presented here. In the general case, it attempts to rearrange the original matrix such that round-off errors due to finite precision arithmetic are reduced as much as possible. The rearrangement of the system matrix A amounts to its being multiplied with a permutation matrix P, which reshuffles its rows; the resulting system that is being solved is then $P A \mathbf{x} = P \mathbf{b}$. For example, for the matrix above, the permuted matrix is

$$PA = \begin{bmatrix} 0 & 1 & 0 \\ 0 & 0 & 1 \\ 1 & 0 & 0 \end{bmatrix} \begin{bmatrix} 1 & 1 & 1 \\ 2 & 3 & 4 \\ 2 & -1 & -5 \end{bmatrix} = \begin{bmatrix} 2 & 3 & 4 \\ 2 & -1 & -5 \\ 1 & 1 & 1 \end{bmatrix},$$

so the actual system that is being solved has what originally was the second equation in the first position, etc. To obtain explicitly the three factors L, U and P of A such that $PA = LU$, one can use the more complete form of the command:

```
>> [L, U, P] = lu(A)
>> P * A == L * U    %checks the factorization.
```

5.1.2 Iterative Methods for $A\mathbf{x} = \mathbf{b}$

When solved by Gauss elimination, problems of the form $A\mathbf{x} = \mathbf{b}$ with A a matrix of order n are computationally intensive, as the cost of the forward elimination stage is $\mathcal{O}(n^3)$. However, remember that multiplying A with a vector only requires $\mathcal{O}(n^2)$. Therefore, if one can solve the system by doing $A \cdot ($ a vector $)$ m times, where m is much smaller than n, one potentially saves a large amount of computing time. This is, in a nutshell, what iterative methods attempt to do. The two such methods that will be presented here are the Jacobi and Gauss-Seidel methods.

5.1.2.1 The Jacobi method

Suppose one wants to solve the system of linear equations $A\mathbf{x} = \mathbf{b}$, which, when written explicitly, becomes:

$$\begin{cases} a_{11}x_1 + a_{12}x_2 + \cdots + a_{1n}x_n = b_1, \\ a_{21}x_1 + a_{22}x_2 + \cdots + a_{2n}x_n = b_2, \\ \quad\vdots \\ a_{n1}x_n + a_{n2}x_2 + \cdots + a_{nn}x_n = b_n. \end{cases}$$

One can start by expressing x_1 from the first equation, x_2 from the second and so forth:

$$x_1 = \frac{b_1 - a_{12}x_2 - a_{13}x_3 - \cdots - a_{1n}x_n}{a_{11}},$$

$$x_2 = \frac{b_2 - a_{21}x_1 - a_{23}x_3 - \cdots - a_{2n}x_n}{a_{22}},$$

$$\vdots$$

$$x_n = \frac{b_n - a_{n1}x_1 - a_{n2}x_2 - \cdots - a_{n,n-1}x_{n-1}}{a_{nn}}.$$

Let us stop for a moment and think about how one could use these relationships in an iterative process. As has been seen in the previous chapter, under such a scenario one starts from some initial guess $\mathbf{x}^0$ and uses this guess, known as the current iterate, to construct the next iterate $\mathbf{x}^1$. The latter then takes the place of the current iterate $\mathbf{x}^0$ and the process is then repeated until convergence is achieved. It is obvious that the current iterate can be used on the right-hand side of the equality signs to define the elements on the left-hand side which will thus naturally become the next iterate. In other words, the equations that define the next iterate from the current iterate can be written, for $k = 0, 1, 2, \ldots$:

$$x_1^{k+1} = \frac{b_1 - a_{12}x_2^k - a_{13}x_3^k - \cdots - a_{1n}x_n^k}{a_{11}},$$

$$x_2^{k+1} = \frac{b_2 - a_{21}x_1^k - a_{23}x_3^k - \cdots - a_{2n}x_n^k}{a_{22}}, \tag{5.1}$$

$$\vdots$$

$$x_n^{k+1} = \frac{b_n - a_{n1}x_1^k - a_{n2}x_2^k - \cdots - a_{n,n-1}x_{n-1}^k}{a_{nn}},$$

or more concisely

$$x_i^{k+1} = \frac{b_i - \sum_{\substack{j=1 \\ j \neq i}}^{n} a_{ij}x_j^k}{a_{ii}}, \quad i = 1, 2, \ldots, n. \tag{5.2}$$

These relationships can be used to update the vector $\mathbf{x}$ in an iterative way. To stop the iterative process, in agreement with the discussion in the previous chapter, one could contemplate at least a couple of ways:

1) Finding, or being close to, a fixed point:

$$\|\mathbf{x}^{k+1} - \mathbf{x}^k\|_2 < T,$$

where

$$\|\mathbf{x}^{k+1} - \mathbf{x}^k\|_2^2 = \sum_{i=1}^{n} \left(x_i^{k+1} - x_i^k\right)^2.$$

2) Driving the residual vector:

$$\mathbf{r}^{k+1} = \mathbf{b} - A\mathbf{x}^{k+1}$$

as close to zero as desired by asking that

$$\|\mathbf{r}^{k+1}\|_2 < \alpha.$$

The 2-norm of a vector, which has been used here, can be easily accessed in MATLAB via the `norm` function without any other arguments.

Although the above relationships are used in a practical implementation, to discuss the convergence of this method is useful to write it in matrix form:

$$\mathbf{x}^{k+1} = D_A^{-1}[(D_A - A)\mathbf{x}^k + \mathbf{b}] = D_A^{-1}[\mathbf{b} - (L_A + U_A)\mathbf{x}^k] = G_J\mathbf{x}^k + \gamma_J,$$

where

$$D_A = \begin{bmatrix} a_{11} & 0 & \cdots & 0 \\ 0 & a_{22} & & \vdots \\ \vdots & & \ddots & 0 \\ 0 & \cdots & 0 & a_{nn} \end{bmatrix} = \text{diagonal of } A,$$

$$L_A = \begin{bmatrix} 0 & & \cdots & & 0 \\ a_{21} & 0 & & & \vdots \\ a_{31} & a_{32} & \ddots & & \\ \vdots & & \ddots & & \\ a_{n1} & \cdots & & a_{n,n-1} & 0 \end{bmatrix} = \text{lower triangular part of } A,$$

$$U_A = \begin{bmatrix} 0 & a_{12} & \cdots & a_{1n} \\ & 0 & a_{23} & \\ & & \ddots & \vdots \\ & & & 0 \end{bmatrix} = \text{upper-triangular part of } A.$$

To see how this comes about, notice that

$$(L_A + U_A)\mathbf{x}^k = \begin{bmatrix} a_{12}x_2^k + a_{13}x_3^k + \ldots + a_{1n}x_n^k \\ a_{21}x_1^k + a_{23}x_3^k + \ldots + a_{2n}x_n^k \\ \vdots \\ a_{n1}x_1^k + a_{n2}x_2^k + \ldots + a_{n,n-1}x_{n-1}^k \end{bmatrix},$$

while

$$D_A^{-1} = \begin{bmatrix} \frac{1}{a_{11}} & 0 & \cdots & 0 \\ 0 & \frac{1}{a_{22}} & & \vdots \\ \vdots & & \ddots & 0 \\ 0 & \cdots & 0 & \frac{1}{a_{nn}} \end{bmatrix},$$

so that multiplication of a column vector by this inverse amounts to dividing all the elements of the vector by the diagonal elements of A.

In this matrix form the term $\gamma_J = D_A^{-1}\mathbf{b}$ brings in the influence of the right-hand side vector $\mathbf{b}$, while $G_J = D_A^{-1}(D_A - A)$ is the convergence matrix for the Jacobi method. This way of writing the Jacobi method makes it clear that one deals with the matrix form of fixed-point iteration,

$$\mathbf{x}^{k+1} = g(\mathbf{x}^k) = G_J\mathbf{x}^k + \gamma_J.$$

To best understand what happens during the iterative process, it is useful to write this relationship for the first few iterates:

$$\mathbf{x}^1 = G_J\mathbf{x}^0 + \gamma_J,$$

$$\mathbf{x}^2 = G_J\mathbf{x}^1 + \gamma_J = G_J\left(G_J\mathbf{x}^0 + \gamma_J\right) + \gamma_J = G_J^2\mathbf{x}^0 + G_J\gamma_J + \gamma_J,$$

$$\vdots$$

$$\mathbf{x}^k = G_J^k\mathbf{x}^0 + \left(G_J^{k-1}\gamma_J + G_J^{k-2}\gamma_J + \ldots\right).$$

As the iteration unfolds, the initial guess $\mathbf{x}^0$ appearing in the first term on the right of the equal sign in the last equation is multiplied by increasingly higher powers of the iteration matrix G_J. The terms in between parantheses in this last equation, on the other hand, depend only on the entries in the system matrix and the right-hand side vector. Since the solution should not depend on the initial guess, the $G_J^k\mathbf{x}^0$ part in the last equation should eventually vanish. Therefore, the Jacobi method converges if and only if G_J^k goes to the zero matrix as $k \to \infty$. In view of the similarity transformations discussed in Chapter 2, assuming the matrix G_J has a full set of eigenvectors, this convergence requirement is equivalent to all the eigenvalues λ_i of the matrix G_J being smaller than one in absolute value. A more convenient way to state this result is by defining the *spectral radius* of this matrix, $\rho(G_J) = \max_i |\lambda_i|$. Using this definition, convergence is equivalent to

$$\rho(G_J) < 1.$$

The matrix G_J is not readily provided and must be computed in order to obtain its spectral radius, as initially one only has access to the system matrix A. For an in-depth convergence investigation, the matrix G_J can be readily constructed in MATLAB. The first part of the script 5.1 shows how this can be done. The script also offers a possible first implementation of the iterative process in equation (5.2) for a fixed value of n.

Notwithstanding this convergence analysis, for all practical purposes one can use a sufficient condition for convergence that only requires the knowledge of the system matrix. The matrix A is defined to be *strictly diagonally dominant* if

$$|a_{ii}| > \sum_{\substack{j=1 \\ j \neq i}}^{n} |a_{ij}|.$$

It can be shown that Jacobi's method always converges if A is strictly diagonally dominant[1]. Notice that having a strictly diagonally dominant system matrix is a sufficient condition for the method to converge, but it is not always necessary. The method may converge even if the system matrix is not strictly diagonally dominant, although there is no guarantee that it will. On the other hand, a considerable drawback of this method is related to the fact that convergence depends on the order of the equations in the system. As an example of what can happen, consider the system

$$\begin{cases} 3x + y + z = 5, \\ x + 5y - z = 5, \\ 2x + y + 4z = 7, \end{cases} \tag{5.3}$$

for which the system matrix is clearly strictly diagonally dominant. Jacobi's method will converge if applied to this system, written in this form. Spelled out explicitly, the updating relationships become:

$$\begin{cases} x^{k+1} = \dfrac{5 - y^k - z^k}{3}, \\ y^{k+1} = \dfrac{5 - x^k + z^k}{5}, \\ z^{k+1} = \dfrac{7 - 2x^k - y^k}{4}. \end{cases} \tag{5.4}$$

On the other hand, the same system can be rearranged as:

$$\begin{cases} x + 5y - z = 5, \\ 3x + y + z = 5, \\ 2x + y + 4z = 7. \end{cases} \tag{5.5}$$

This system clearly has the same solution as the one just above, but the matrix is not strictly diagonally dominant and Jacobi's method is not guaranteed to converge. Therefore, one needs to have a good knowledge of the problem to be solved before using this method. The good news, however, is that a large array of engineering problems lead to strictly diagonally dominant matrices when approximated by the finite element method. The latter is a first choice nowadays for the numerical solution of partial differential equations and is widely used in technological design processes.

---

[1]For a strictly diagonally dominant matrix, we have $\|G_J\|_\infty = \left\|D_A^{-1}(D_A - A)\right\|_\infty = \max\limits_i \sum\limits_{j \neq i} \dfrac{|a_{ij}|}{|a_{ii}|} < 1$. The matrix infinity norm is *equivalent to* the matrix 2-norm, which is equivalent to the spectral radius of G_J.

MATLAB Script 5.1: Spectral Radius, Jacobi

```
clear all;        % clears all variables
A = [ 3, 1, 1; 1, 5, -1; -2, 1, 7]; %example matrix
Da = diag( diag(A) ); %diagonal entries of A
La = tril(A) - Da;  % lower-diagonal entries of A
Ua = triu(A) - Da; % upper-diagonal entries of A
GJ = - inv(Da) * ( Ua + La);
rhoGJ = max( abs( eig(GJ) )  )
% now start Jacobi iteration
b = [ 5; 4; 9];
x0 = [ 0; 0; 0];
x1 = x0; % to make x1 same type/size as x0
for k = 0:100
    % update all components
    for i=1:3
        s = 0;
        % a more efficient way to compute the sum
        for j = 1:3
            if i ~= j
                s = s + A(i,j)*x0(j);
            end     % if
        end         % for
        x1(i) = ( b(i) - s ) / A(i,i);
    end
    res = A * x1 - b;
    if norm( res ) < 10^(-5)
        break;
    end
    x0 = x1;
end
x1 % prints the solution
```

5.1.2.2   Computational efficiency

The script 5.1 above, which attempts a direct translation of equation (5.2) into code, is not the most efficient way to implement the method on current computers. The reason is the `if` statement that is imbedded in three loops. The script is written for $n = 3$ equations, but let us consider what happens if the number of equations becomes large. This `if` statement is executed n^2 times for every iteration (that is, each time the index k is incremented) and as the dimension of the system grows, $n \to \infty$, the computational cost can easily become exhaustive. Moreover, this is not a price that we need to pay: the loop on the j index can easily be broken up in two loops, as the script below shows. It is good to remember that, while today's computers have large amounts of memory and tremendous processing power, it is still a fact that most problems in computational science exploit these resources at the maximum. Therefore, a successful computational scientist needs to be aware of all possible ways a computer code can be made more efficient. The MATLAB script 5.2 shows this more efficient implementation. Several other changes have also been incorporated to make the code more general.

MATLAB Script 5.2: More efficient Jacobi implementation

```
clear all;        % clears all variables
A = [ 3, 1, 1; 1, 5, -1; -2, 1, 7]; %example matrix
% start Jacobi iteration
b = [ 5; 4; 9];
x0 = [ 0; 0; 0];
x1 = x0; % to make x1 same type as x0
maxIterations = 100;
for k = 0:maxIterations
    % update all components
    for i=1:n
        s = 0;
        % a more efficient way to compute the sum
        for j = 1:(i-1)
            s = s + A(i,j)*x0(j);
        end
        for j = (i+1):n
            s = s + A(i,j)*x0(j);
        end
        x1(i) = ( b(i) - s ) / A(i,i);
    end
    res = A * x1 - b;
    if norm( res ) < 10^(-5)
        break;
    end
    x0 = x1;
end
x1 % prints the solution
```

5.1.2.3   The Gauss-Seidel method

The Gauss-Seidel method is based on a rather straightforward observation. Notice that, in the Jacobi iteration relationships (5.1)–(5.2), once the value of x_1^{k+1} is evaluated from the first equation, one can already use it on the right-hand side for x_2^{k+1}. Continuing in this manner, after the first two unknowns have been updated, their newly found values can be used on the right-hand side for the third unknown, and so forth. In general, this would clearly amount to using more up-to-date values and thus possibly faster convergence. The Gauss-Seidel method exploits this idea. The update rules corresponding to 5.4 take the form:

$$
\begin{cases}
x^{k+1} = \dfrac{5 - y^k - z^k}{3}, & \\[2mm]
y^{k+1} = \dfrac{5 - x^{k+1} + z^k}{5}; & \text{use } x^{k+1} \text{ here!} \\[2mm]
z^{k+1} = \dfrac{7 - 2x^{k+1} - y^{k+1}}{4}; & \text{use } x^{k+1}, y^{k+1}!
\end{cases}
\tag{5.6}
$$

For the case of a more general system, the Gauss-Seidel updating relationships are:

$$x_1^{k+1} = \frac{b_1 - a_{12}x_2^k - a_{13}x_3^k - \cdots - a_{1n}x_n^k}{a_{11}},$$

$$x_2^{k+1} = \frac{b_2 - a_{21}x_1^{k+1} - a_{23}x_3^k - \cdots - a_{2n}x_n^k}{a_{22}}, \qquad (5.7)$$

$$\vdots$$

$$x_n^{k+1} = \frac{b_n - a_{n1}x_1^{k+1} - a_{n2}x_2^{k+1} - \cdots - a_{n,n-1}x_{n-1}^{k+1}}{a_{nn}},$$

or more concisely

$$x_i^{k+1} = \frac{b_i - \sum_{j=1}^{i-1} a_{ij}x_j^{k+1} - \sum_{j=i+1}^{n} a_{ij}x_j^k}{a_{ii}}, \quad i = 1, 2, \ldots, n. \qquad (5.8)$$

The reader is invited to show that the Gauss-Seidel method can be written in matrix form as:

$$\mathbf{x}^{k+1} = -(D_A + L_A)^{-1}U_A \mathbf{x}^k + (D_A + L_A)^{-1}\mathbf{b} = G_S\mathbf{x}^k + \gamma_S,$$

which again represents a relationship of fixed-point iteration type. Just like the Jacobi method, a strictly diagonally dominant matrix is a sufficient condition for the Gauss-Seidel method to converge. The method does converge faster than Jacobi's method in most cases, but this is not guaranteed. In fact, there are problems for which either one of them converges but the other does not. As a side note for the classical Jacobi and Gauss-Seidel methods, which cannot be elaborated here in all detail, it is still worth mentioning that the Jacobi method can more easily take advantage of parallel processing than the Gauss-Seidel method. Because of this feature, the Jacobi method is still in use in modern large-scale problems.

Finally, we mention that more powerful methods are routinely used nowadays. One such method is the conjugate gradient method, which will be presented in section 8.4.3. The conjugate gradient method, related to the steepest descent method which is used for optimization problems, is a member of a family of modern and effective methods, known as Krylov subspace methods, for solving large-scale linear systems.

5.2 Newton's Method for Nonlinear Systems

It should not be a surprise that nonlinear systems arise just so often in applications as do their linear counterparts. As a simple example, finding the possible points of intersection of three non-planar surfaces, as opposed to the intersection of three planes, is equivalent to solving three nonlinear equations for the three coordinates of the intersection.

In the most general case, one is interested in solving a system of n equations with n unknowns:

$$\begin{cases} f_1(x_1, x_2, \ldots, x_n) = 0, \\ f_2(x_1, x_2, \ldots, x_n) = 0, \\ \quad\vdots \\ f_N(x_1, x_2, \ldots, x_n) = 0. \end{cases}$$

The focus will again be on finding an iterative solution procedure that starts with some initial guess,

$$\begin{pmatrix} x_1^0 \\ x_2^0 \\ \vdots \\ x_n^0 \end{pmatrix} = \mathbf{x}^0 = \text{ given.}$$

In view of the approach used in Newton's method for one equation, one may consider using the Taylor series for the nonlinear functions $f_1, f_2, \ldots, f_n$ around the point $\mathbf{x}^0$ and keeping only the linear terms. This is also known as *linearization*; the function values at some other point $\mathbf{x}^1$, supposed to be close to $\mathbf{x}^0$, will be given by:

$$\begin{cases} f_1(\mathbf{x}^1) = f_1(\mathbf{x}^0) + \vec{\nabla} f_1|_{\mathbf{x}^0} \cdot (\mathbf{x}^1 - \mathbf{x}^0) + \text{ H.O.T}, \\ f_2(\mathbf{x}^1) = f_2(\mathbf{x}^0) + \vec{\nabla} f_2|_{\mathbf{x}^0} \cdot (\mathbf{x}^1 - \mathbf{x}^0) + \text{ H.O.T}, \\ \quad\vdots \\ f_n(\mathbf{x}^1) = f_n(\mathbf{x}^0) + \vec{\nabla} f_n|_{\mathbf{x}^0} \cdot (\mathbf{x}^1 - \mathbf{x}^0) + \text{ H.O.T}, \end{cases}$$

where

$$\vec{\nabla} f_i|_{\mathbf{x}^0} = \left[\frac{\partial f_i}{\partial x_1} \quad \frac{\partial f_i}{\partial x_2} \quad \cdots \quad \frac{\partial f_i}{\partial x_N} \right]_{\mathbf{x}^0}$$

is a row-vector of all the first partial derivatives for a given function f_i, computed at $\mathbf{x}^0$. Proceeding again as in the scalar Newton's method by asking that $f_1(\mathbf{x}^1) = 0$, $f_2(\mathbf{x}^1) = 0$ and so on, one obtains:

$$0 = f_1(\mathbf{x}^0) + \vec{\nabla} f_1|_{\mathbf{x}^0} \cdot (\mathbf{x}^1 - \mathbf{x}^0),$$
$$0 = f_2(\mathbf{x}^0) + \vec{\nabla} f_2|_{\mathbf{x}^0} \cdot (\mathbf{x}^1 - \mathbf{x}^0),$$
$$\vdots$$
$$0 = f_n(\mathbf{x}^0) + \vec{\nabla} f_n|_{\mathbf{x}^0} \cdot (\mathbf{x}^1 - \mathbf{x}^0).$$

Conveniently rearranging the terms leads to:

$$-\begin{bmatrix} f_1(\mathbf{x}^0) \\ f_2(\mathbf{x}^0) \\ \vdots \\ f_n(\mathbf{x}^0) \end{bmatrix} = J|_{\mathbf{x}^0}\left(\mathbf{x}^1 - \mathbf{x}^0\right),$$

where $J|_{\mathbf{x}^0}$ is a matrix which has its rows equal to the gradient vectors:

$$J|_{\mathbf{x}^0} = \begin{bmatrix} \vec{\nabla}f_1|_{\mathbf{x}^0} \\ \vec{\nabla}f_2|_{\mathbf{x}^0} \\ \vdots \\ \vec{\nabla}f_n|_{\mathbf{x}^0} \end{bmatrix} = \begin{bmatrix} \frac{\partial f_1}{\partial x_1} & \frac{\partial f_1}{\partial x_2} & \cdots & \frac{\partial f_1}{\partial x_n} \\ \frac{\partial f_2}{\partial x_1} & \frac{\partial f_2}{\partial x_2} & \cdots & \frac{\partial f_2}{\partial x_n} \\ \vdots & \vdots & \vdots & \vdots \\ \frac{\partial f_n}{\partial x_1} & \frac{\partial f_n}{\partial x_2} & \cdots & \frac{\partial f_n}{\partial x_n} \end{bmatrix},$$

known as the *Jacobian matrix* of $f_1, f_2, \ldots$, with respect to the n variables $x_1, x_2, \ldots$, at the point $\mathbf{x}^0$. Furthermore, notice that the quantity

$$\begin{bmatrix} f_1(\mathbf{x}^0) \\ f_2(\mathbf{x}^0) \\ \vdots \\ f_n(\mathbf{x}^0) \end{bmatrix} = \mathbf{r}^0$$

is an easy-to-evaluate vector of residual values of the same n functions at $\mathbf{x}^0$. The above relationship becomes

$$J|_{\mathbf{x}^0}\left(\mathbf{x}^1 - \mathbf{x}^0\right) = -\mathbf{r}^0, \tag{5.9}$$

which can be seen to represent a system of equations. Given a non-singular Jacobian matrix, the system can be solved and thus the difference vector $\mathbf{x}^1 - \mathbf{x}^0 = \vec{\triangle}^1$ computed. Then the new approximation to the solution is obtained by updating the current iterate, using $\mathbf{x}^1 = \mathbf{x}^0 + \vec{\triangle}^1$. This new iterate $\mathbf{x}^1$ replaces the previous iterate $\mathbf{x}^0$ and the process is repeated until convergence. That is, for $k = 0, 1, \ldots$ the iterative process constructs the sequence of iterates:

$$\mathbf{x}^{k+1} = \mathbf{x}^k + \vec{\triangle}^{k+1}, \quad \text{where } J|_{\mathbf{x}^k}\vec{\triangle}^{k+1} = -\mathbf{r}^k.$$

5.2.1 An Example with Newton's Method for a System

Consider the following system of three nonlinear equations with three unknowns:

$$\begin{bmatrix} f_1(x, y, z) \\ f_2(x, y, z) \\ f_3(x, y, z) \end{bmatrix} = \begin{bmatrix} x^2 + y^2 + z^2 - 1 \\ x^2 + y^2 - z^2 \\ x - y \end{bmatrix} = \mathbf{0}. \tag{5.10}$$

It represents the intersection of three surfaces: a sphere with radius one centered at the origin, a cone with the apex at the origin and a plane whose trace in the xy-plane is the first bisectrix. The system is interesting because not only can it be solved analytically but it is also easy to visualize the solutions, which allows us to investigate some of the issues that may arise when tackling this kind of problems computationally. The sphere will intersect the cone in two circles, one located above the xy-plane, at $z = \sqrt{2}/2$, the other below the xy-plane at $z = -\sqrt{2}/2$. The plane $x = y$ will intersect each circle in two points, so there are four solutions to this system. Choosing a different plane may lead to very different scenarios, from no solutions (for example if the plane were $z = 1$ instead of $x = y$) to an infinity of solutions (for the planes $z = \pm\sqrt{2}/2$) with cases when there are one, two, three and four solutions in between.

With the above discussion in mind, it is obvious that the solution which Newton's method may find will depend on the initial guess. For example, starting with the initial guess $\begin{bmatrix} x \\ y \\ z \end{bmatrix}^0 = \begin{bmatrix} 0.1 \\ 0.2 \\ 0.3 \end{bmatrix}$, Newton's method will converge to the solution $x = y = 0.5$, $z = \sqrt{2}/2$. The first step of Newton's Method entails the following:

(1) Setting up the Jacobian Matrix; for this case, the derivatives are:

$$\frac{\partial f_1}{\partial x} = 2x \qquad \frac{\partial f_1}{\partial y} = 2y \qquad \frac{\partial f_1}{\partial z} = 2z$$

$$\frac{\partial f_2}{\partial x} = 2x \qquad \frac{\partial f_2}{\partial y} = 2y \qquad \frac{\partial f_2}{\partial z} = -2z$$

$$\frac{\partial f_3}{\partial x} = 1 \qquad \frac{\partial f_3}{\partial y} = -1 \qquad \frac{\partial f_3}{\partial z} = 0$$

Thus, it follows that at the point x^0 the Jacobian matrix is:

$$J^0 = \begin{bmatrix} 0.2 & 0.4 & 0.6 \\ 0.2 & 0.4 & -0.6 \\ 1 & -1 & 0 \end{bmatrix}.$$

(2) Evaluating the residual vector:

$$r^0 = f(x^0) = \begin{bmatrix} -0.86 \\ -0.0399 \\ -0.1 \end{bmatrix}$$

with $\|r^0\| = 0.866718$.

(3) Computing the increment $x^1 - x^0 = \vec{\Delta}^1$, which is a solution to the system

$$[J^0] \cdot \vec{\Delta}^1 = -r^0.$$

Using the notation:

$$\vec{\triangle}^1 = \begin{bmatrix} \delta_1^1 \\ \delta_2^1 \\ \delta_3^1 \end{bmatrix},$$

the solution of the system, which can readily be obtained using MAT-LAB, is: $\delta_1^1 = 0.81666667$, $\delta_2^1 = 0.71666667$ and $\delta_3^1 = 0.68333333$.

(4) Obtaining the new position (the new approximation to the solution):

$$\mathbf{x}^1 = \begin{bmatrix} 0.1 \\ 0.2 \\ 0.3 \end{bmatrix} + \vec{\triangle}^1 = \begin{bmatrix} 0.91666667 \\ 0.91666667 \\ 0.98333333 \end{bmatrix}.$$

The same calculations are now repeated with the new value for $\mathbf{x}$; while further computations will not be shown, it is interesting to examine the evolution of the residual over just one time step for this system. The new residual vector is:

$$\mathbf{r}^1 = \mathbf{f}(\mathbf{x}^1) = \begin{bmatrix} 1.64749999999 \\ 0.7136111111 \\ -1.110223e - 16 \end{bmatrix},$$

with norm $\|\mathbf{r}^1\| = 1.7954099999446456$. The norm of the residual increases over the first iteration instead of decreasing, which may come as a surprise for such a simple system. Nevertheless, one more iteration decreases the residual norm more than five times; the solution converges to an accuracy of 10^{-8} in only six iterations.

5.3 MATLAB® Built-in Functions

The MATLAB function `fsolve` solves systems of nonlinear equations. It implements a set of three algorithms. The default is a trust-region dogleg method, which was not discussed in this text; nevertheless, the use of the function is rather straightforward. To do so, one needs to first prepare the system to be solved as shown in the code excerpt below. In this example, `fsolve` is not instructed to output more than the minimal information.

MATLAB Script 5.3: Using fsolve for a system

```
clear all;        % clears all variables
x0 = [0.1, 0.2, 0.3];
xStar = fsolve( 'myFun', x0 ) % output: 0.5000  0.5000  0.7071

% This should be in file myFun.m
function [val] = myFun( x )
val(1) = x(1)^2 + x(2)^2 + x(3)^2 - 1.0;
val(2) = x(1)^2 + x(2)^2 - x(3)^2;
val(3) = x(1) - x(2);
end % function
```

In this default call, the number of iterations is set to four hundred and the Jacobian matrix is not needed. When the Jacobian is available, it can be provided as a second output from the user-defined function. The next example script 5.4 shows how to change the default behavior of `fsolve` to provide a table with information for each iteration and also use the Jacobian by sending optional input parameters. The advantage of providing the Jacobian should not be underestimated: this new version only uses seven function evaluations (iterations), as opposed to twenty five evaluations in the default case.

MATLAB Script 5.4: Using fsolve with provided Jacobian

```
clear all;        % clears all variables
x0 = [0.1, 0.2, 0.3];
options = optimoptions('fsolve','Display','iter',...
    'SpecifyObjectiveGradient',true);
xStar = fsolve( 'myFun', x0, options )

function [val, J] = myFun (x)

val(1) = x(1)^2 + x(2)^2 + x(3)^2 - 1.0;
val(2) = x(1)^2 + x(2)^2 - x(3)^2;
val(3) = x(1) - x(2);
J(1,1) = 2 * x(1);
J(1,2) = 2 * x(2);
J(1,3) = 2 * x(3);
J(2,1) = 2 * x(1);
J(2,2) = 2 * x(2);
J(2,3) = - 2 * x(3);
J(3,1) = 1.0 ;
J(3,2) = -1.0;
J(3,3) = 0.0;

end
```

5.4 Exercises

Ex. 1 Solve the twenty-one systems of equations:

$$\begin{cases} 5x + y + z = 0.01m^2 - 2m \\ x - 5y + 2z = 1 - m \\ 2x + y + 4z = 10 \end{cases}$$

obtained by setting, in turn, the value of the quantity m on the right-hand side to all integers in between, and including, $m = 0$ and $m = 20$. Use the LU-decomposition of the system matrix that is implemented in MATLAB by the **lu** function. Perform the decomposition only once, then use the lower- and upper-triangular factors repeatedly to find each successive solution. Plot the first and second components of the solution (that is, the values of x and y) as a function of the right-hand side parameter m.

Ex. 2 Solve, with a MATLAB code of your own, the following system:

$$\begin{cases} 5x - y + z + w = 7 \\ x + 7y + 2z + 2w = 5 \\ 2x + y + 5z + w = 8 \\ x - y + z + 4w = 6 \end{cases}$$

Use the Jacobi method with a tolerance of 10^{-6} for the norm of the residual. Arrange your results in a table of the form:

iteration	x	y	z	w
0	0	0	0	0
1	...	...	...	...
⋮				

so that you can see how x, y, z and w change with each iteration. Remember that you can create such a table using **disp()** in a loop. For an even better-looking table, you may also use **fprintf**. As obvious from the table above, start with the zero guess. Moreover, plot the norm of the residual versus the iteration number, using a logarithmic scale on the vertical axis (the residual axis). The code can be organized as: (a) a function file that implements Jacobi's method and (b) a script that calls the function with the appropriate inputs and processes the results.

Ex. 3 Solve again the system above, in MATLAB, using the Gauss-Seidel method. Produce the same results as for the previous problem.

Ex. 4 Consider the system of linear equation $Ax = b$, where the matrix A is given by:

$$A = \begin{pmatrix} 1.1 & 1 & 1 \\ 1 & 1.1 & 1 \\ 1 & 1 & 1.1 \end{pmatrix}$$

Compute its spectral radius $\rho(G_J)$ as shown in MATLAB script 5.1 for the Jacobi method. You will find that, for this matrix, $\rho(G_J) > 1$. Use your own Jacobi method MATLAB code to attempt solving a linear system with this matrix. To create such a system, generate what will be the true solution by creating a random vector with the MATLAB command: x=rand(3,1). The right-hand side will then be by the product $b = Ax$. Now solve $Ax = b$ for x. Let the initial guess $x^0 = [0, 0, 0]^T$, the maximum number of iterations be 1000, and the accuracy $\alpha = 1e{-}12$; so ask that $\|b - Ax^{k+1}\|_2 < \alpha$. What do you observe? Does the Jacobi method converge?

Ex. 5 Repeat the same calculation as in **Ex. 4** for the Gauss-Seidel method. Is $\rho(G_S) > 1$? Does the Gauss-Seidel method converge?

Ex. 6 Consider the below system of linear equations $Ax = b$

$$\begin{pmatrix} 2 & -1 & 0 & 0 & 0 & 0 & 0 & 0 \\ -1 & 2 & -1 & 0 & 0 & 0 & 0 & 0 \\ 0 & -1 & 2 & -1 & 0 & 0 & 0 & 0 \\ 0 & 0 & -1 & 2 & -1 & 0 & 0 & 0 \\ 0 & 0 & 0 & -1 & 2 & -1 & 0 & 0 \\ 0 & 0 & 0 & 0 & -1 & 2 & -1 & 0 \\ 0 & 0 & 0 & 0 & 0 & -1 & 2 & -1 \\ 0 & 0 & 0 & 0 & 0 & 0 & -1 & 2 \end{pmatrix} \begin{pmatrix} x_1 \\ x_2 \\ x_3 \\ x_4 \\ x_5 \\ x_6 \\ x_7 \\ x_8 \end{pmatrix} = \begin{pmatrix} 1 \\ 0 \\ 0 \\ 0 \\ 0 \\ 0 \\ 0 \\ 1 \end{pmatrix}.$$

One can show that the system matrix A is symmetric and positive definite, but clearly it is not strictly diagonally dominant. The true solution of this linear system is $x = [1, 1, 1, 1, 1, 1, 1, 1]^T$.

(a) Use the Jacobi and the Gauss-Seidel methods with MATLAB to solve the above linear system. Use the initial guess $x^0 = [0, 0, 0, 0, 0, 0, 0, 0]^T$ and the residual tolerance $\alpha = 1e{-}12$, where $\|b - Ax^{k+1}\|_2 < \alpha$. Observe that both methods converges, despite the fact that the matrix A is not strictly diagonally dominant. What is the number of iterations for each method? Compute the true errors by $\|x - x^{k+1}\|_2$ for both methods. Are they at the same order as α?

(b) Notice that the equations for the *odd-indexed* unknowns only involve the *even-indexed* unknowns, not other odd-indexed unknowns, and vice versa. One can thus group the odd-indexed and even-indexed unknowns separately to obtain the following system of equations;

$$
\begin{pmatrix}
2 & 0 & 0 & 0 & -1 & 0 & 0 & 0 \\
0 & 2 & 0 & 0 & -1 & -1 & 0 & 0 \\
0 & 0 & 2 & 0 & 0 & -1 & -1 & 0 \\
0 & 0 & 0 & 2 & 0 & 0 & -1 & -1 \\
-1 & -1 & 0 & 0 & 2 & 0 & 0 & 0 \\
0 & -1 & -1 & 0 & 0 & 2 & 0 & 0 \\
0 & 0 & -1 & -1 & 0 & 0 & 2 & 0 \\
0 & 0 & 0 & -1 & 0 & 0 & 0 & 2
\end{pmatrix}
\begin{pmatrix}
x_1 \\ x_3 \\ x_5 \\ x_7 \\ x_2 \\ x_4 \\ x_6 \\ x_8
\end{pmatrix}
=
\begin{pmatrix}
1 \\ 0 \\ 0 \\ 0 \\ 0 \\ 0 \\ 0 \\ 1
\end{pmatrix}.
$$

The new ordering of unknowns renders a general algorithm that allows one to do what can be thought of as two decoupled Jacobi half-sweeps, each one updating half of the unknowns. They are described by the following pseudocode:

for $i = 1 : 2 : n$

$$
x_i^{k+1} = \frac{1}{a_{ii}} \left[-\sum_{\substack{j=1 \\ j \neq i}}^{n} a_{ij} x_j^k + b_i \right]
$$

end

for $i = 2 : 2 : n$

$$
x_i^{k+1} = \frac{1}{a_{ii}} \left[-\sum_{\substack{j=1 \\ j \neq i}}^{n} a_{ij} x_j^{k+1} + b_i \right]
$$

end

This is known as the *red-black Gauss-Seidel* method. Can you see why it may have an advantage over the original Jacobi method? Implement the red-black Gauss-Seidel method in MATLAB to solve this linear system. What is the number of iterations for the method to converge? Compute the true error.

Ex. 7 Do by hand, on paper, one iteration of Newton's method for the nonlinear system:

$$
\begin{cases}
x^2 + y^3 - 1 = 0 \\
x^3 - y^2 + 0.25 = 0.
\end{cases}
$$

Start with the initial guess $\vec{x}^0 = [x^0, \ y^0]^T = [0.5, \ 0.5]^T$ and compute the next iterate $\vec{x}^1$. Also compare the norm of the residual at the new iterate with the residual norm computed for the initial guess.

Ex. 8 Solve the system

$$\begin{cases} x + y^2 + z - 4 = 0 \\ x^2 + y + z - 6 = 0 \\ x + y + z^2 - 4 = 0 \end{cases}$$

using a residual tolerance of 10^{-6} and an initial guess of your choice.

Ex. 9 Write a MATLAB script for solving the following system of nonlinear equations by Newton's method

$$2x - 3y + z = 4,$$
$$2x + y - z = -4,$$
$$x^2 + y^2 + z^2 = 4.$$

Let the initial guess be $\mathbf{x}^0 = [x^0, y^0, z^0]^T = [-0.5, 1.5, 1.5]^T$ and use a residual tolerance of 10^{-6}.

Ex. 10 Modify MATLAB scripts 5.3 and 5.4 to solve **Ex. 9**. Compare the results of the two different ways of calling the MATLAB default function with your Newton's method in **Ex. 9.**, in terms of convergence criterion, number of iterations, CPU time, and accuracy.

Chapter 6

Approximation of Functions

The approximation of functions that are only known through a data set of pointwise values is of central importance in scientific computation. It allows us to obtain values in between the data points, to extract general features such as extrema, and lays the foundation for the numerical approximation of derivatives and integrals. The approximation problem may arise because only noisy data about a function are available. It may also be due to the fact that a presumably continuous function can only be characterized by its values at a finite set of points, usually known as a *discretization*. This chapter presents the basic methods for approximation of functions by polynomials and other well-known mathematical entities that can be easily handled numerically.

6.1 A Hypothetical Example

Data obtained through experiments are oftentimes noisy. Nevertheless, such data must still be routinely analyzed in order to reveal the underlying trends. Methods that achieve this by first creating functions that approximate the data, in some well-defined sense, are generally known as *curve fitting* methods.

Suppose for example that a variable voltage source is used to estimate the resistance in an electrical circuit indirectly by measuring the current at some point in the circuit. Furthermore, suppose that the device that measures the current is an old-school analogue amperemeter with a needle turning against a circular scale. Readings are thus not very accurate, and the following table of values is obtained for a series of voltage inputs:

Ohm's law asserts that $V = RI$, but notice that choosing several different couples of points in the table will give different values for R. Clearly, this is due to the inaccuracy in taking the readings for I. What is then the best estimate for the resistance R, given the data in the table, if one considers that each and every one of the data points can be trusted just as much as any other?

One way to address this question, which can be easily extended to more general scenarios, is to define an error for each measurement point. This error

DOI: 10.1201/9780429262876-6

is initially unknown and has the form:

$$E_k = f(V_k) - I_k,$$

where $f(V) = V/R$ is an unknown function, in that it contains the value of the resistance R that has to be determined. For easier notation, the pointwise errors can be naturally collected together in an error vector,

$$\mathbf{E} = \begin{bmatrix} E_1 \\ E_2 \\ E_3 \\ E_4 \end{bmatrix}.$$

Let us now define the constant $A = 1/R$, so that A is the new unknown quantity, and consider the sum of the squares, which is also the square of the norm of the error vector $||\mathbf{E}||^2$, of all these pointwise errors:

$$||\mathbf{E}||^2 = (AV_1 - I_1)^2 + (AV_2 - I_2)^2 + (AV_3 - I_3)^2 + (AV_4 - I_4)^2.$$

One wants to achieve a minimum of this quantity with respect to A. For this end, its derivative, which can be seen to be:

$$\frac{d||\mathbf{E}||^2}{dA} = 2V_1(AV_1 - I_1) + 2V_2(AV_2 - I_2) + 2V_3(AV_3 - I_3) + 2V_4(AV_4 - I_4),$$

is set to zero. This leads to the following equation:

$$A(V_1^2 + V_2^2 + V_3^2 + V_4^2) = V_1I_1 + V_2I_2 + V_3I_3 + V_4I_4,$$

whence the value of the constant, $A = 0.4794$, and consequently $R = 2.086$ can be obtained.

Known as the least-squares technique, this method has the advantage that the solution of the equation, A, will always be a minimum point for the function $||E||^2$. It will be explored in more detail in the next sections.

6.1.1   Least-Squares Fitting Methods

Let us start by introducing a more general and convenient mathematical notation. Consider that a set of n data points, representing the uncertain measurements, is given:

$$\{(x_1, y_1), (x_2, y_2), \ldots, (x_n, y_n)\}.$$

This is the underlying *data set*. Also consider that the owner of the data set has valid knowledge of the fact that the data must obey some law, with deviations from such law only due to measurement errors. For example, the expected law expressing the dependence of y on x could be linear, or it could be exponential, and so on.

Our first goal is to see how to fit a straight line of the form $y = f(x) = Ax + B$ through these data points. This will be done in such a way that a reasonably-defined measure of the deviation from the data points is minimized. In other words, one attempts to find the coefficients A and B that minimize a suitably defined global error. Notice that the data value available for y_k at the point x_k is expected to be off by some error:

$$f(x_k) = Ax_k + B = y_k + E_k \approx y_k,$$

with $f(x_k) - y_k = E_k \neq 0$ in general. Suppose that one tries to minimize the sum of all these errors, $\sum_k E_k$. It can easily be seen that positive errors may counterbalance negative errors in this case, so one could end up with large errors at different points, that nevertheless cancel in the sum. Thus, minimizing just a plain sum of the errors is not a good strategy.

In light of this, consider the error vector defined by:

$$\mathbf{E} = \begin{bmatrix} E_1 \\ E_2 \\ \vdots \\ E_n \end{bmatrix}.$$

Ideally, this vector should have all components equal to zero. Since that may not be the case, the best one can do is to shrink it as much as possible. Several scenarios are possible:

(1) Minimize the maximum component of the error vector (also known as minimizing the L_∞-norm of the error)

$$||\mathbf{E}||_\infty = \max_{1 \leq k \leq n} E_k.$$

(2) Minimize the average error, or the L_1-norm:

$$||\mathbf{E}||_1 = \frac{1}{n} \sum_{k=1}^{n} |E|_k,$$

though the $\frac{1}{n}$ factor need not be included explicitly.

(3) Minimize the distance from the tip of the error vector to the origin (or the L_2-norm of the error):

$$||\mathbf{E}||_2 = \left(\sum_{k=1}^{n} E_k^2 \right)^{\frac{1}{2}}.$$

In practice the last method is the easiest to work with because of intricacies related to taking derivatives in the other two cases. Notice that, due to the monotonicity of the square root function, one can minimize instead the quantity:

$$||\mathbf{E}||_2^2 = \sum_{k=1}^{n} (f(x_k) - y_k)^2 = \sum_{k=1}^{n} (Ax_k + B - y_k)^2.$$

The variables here are the two unknown coefficients A and B, hence the following two equations must be satisfied for an extremum:

$$\frac{\partial ||\mathbf{E}||_2^2}{\partial A} = 0 \implies \sum_{k=1}^{n} 2(Ax_k + B - y_k)x_k = 0$$

$$\frac{\partial ||\mathbf{E}||_2^2}{\partial B} = 0 \implies \sum_{k=1}^{n} 2(Ax_k + B - y_k) = 0$$

Written in system form, these two equations become:

$$\begin{pmatrix} \sum_{k=1}^{n} x_k^2 & \sum_{k=1}^{n} x_k \\ \sum_{k=1}^{n} x_k & n \end{pmatrix} \begin{pmatrix} A \\ B \end{pmatrix} = \begin{pmatrix} \sum_{k=1}^{n} x_k y_k \\ \sum_{k=1}^{n} y_k \end{pmatrix}.$$

This system of equations can be easily set up in MATLAB and solved, using the backslash command, for the values of A and B. Suppose that, instead of a fit of the form $I = AV$ as in the previous section, a fit of the form $V = AI + B$ were searched for the data in Table 6.1, i.e. $V \leftrightarrow y$ and $I \leftrightarrow x$. In that case, the code in script 6.1 could be used to perform the computation and would result in figure 6.1.

MATLAB Script 6.1: Linear Fit

```
x = [0 2.6 4.95 7.02];
y = [0 5 10 15];
x2 = x .* x;
M = [ sum(x2) sum(x) ; ...
      sum(x) length(x)]        % the matrix
xy = x .* y;
rhs = [sum(xy) ; sum(y)];
coeff = M \ rhs;        % solve the system
A = coeff(1)   % A = 2.1304
B = coeff(2)   % B = -0.25989
% Alternate computation
xbar = sum(x) / length(x);
ybar = sum(y) / length(y);
xmn = x - xbar
ymn = y - ybar
A = sum(xmn .* ymn) / sum ( xmn .* xmn)
B = ybar - A * xbar
```

A conceptually different approach to the least-squares minimization problem for a linear fit of the form $y = Ax + B$ is known in statistics as *simple*

TABLE 6.1: Voltage Inputs.

Measurement (k)	1	2	3	4
Voltage (V_k)	0	5	10	15
Current (I_k)	0	2.6	4.95	7.02

linear regression [12, 14]. Using a common statistics notation, the coefficients A and B can be shown to be:

$$A = \frac{\sum_{k=1}^{n}(x_k - \bar{x})(y_k - \bar{y})}{\sum_{k=1}^{n}(x_k - \bar{x})^2}; \quad B = \bar{y} - A\bar{x}$$

where the bar denotes the average over all data points, that is $\bar{x} = (\sum x_k)/n$ and $\bar{y} = (\sum y_k)/n$.

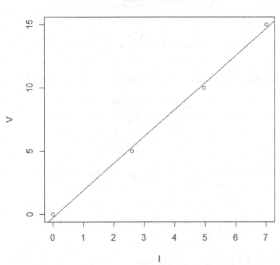

Line of Best Fit

Figure 6.1: Data in table 6.1 plotted together with best straight line fit.

The least-squares method presented above can be easily extended to fits that use higher degree polynomials. For example, to fit a quadratic to a set of data points $\{(x_i, y_i)\}$, $i = 1, 2, \ldots, n$, let the quadratic be given by $f(x) = Ax^2 + Bx + C$.

The error at x_k is

$$E_k = f(x_k) - y_k = Ax_k^2 + Bx_k + C - y_k,$$

and hence asking again for the minimum of the (square of) the L_2-norm of the error boils down to:

$$\frac{\partial\left(\sum_{k=1}^{n} E_k^2\right)}{\partial A} = \frac{\partial\left(\sum_{k=1}^{n} E_k^2\right)}{\partial B} = \frac{\partial\left(\sum_{k=1}^{n} E_k^2\right)}{\partial C} = 0. \tag{6.1}$$

This leads to a 3×3 system of linear equations for the three coefficients A, B, C. Once these coefficients are obtained, the fit is obviously defined and can be used to predict results at various x locations. In closing this section, a mention should be made about the linear systems obtained from the least-squares method. These systems quickly become prone to numerical errors as their dimension increases, therefore extensions of this methodology to relatively high-degree polynomials must be handled with care and may not be desirable.

6.1.2 Fitting with Non-Polynomial Functions

Unfortunately, the method presented so far does not easily extend to fits that are not of the polynomial type. As an example, consider the problem of fitting an exponential function of the form $f(x) = Ce^{Ax}$ to a set of data points (x_k, y_k), $k = 1, 2, \ldots, n$. The error to be minimized is:

$$\|\mathbf{E}\|_2^2 = F(A, C) = \sum_{k=1}^{n} \left(Ce^{Ax_k} - y_k\right)^2 .$$

For the minimum to occur, one therefore needs to ask that

$$\frac{\partial \|\mathbf{E}\|_2^2}{\partial A} = \frac{\partial \|\mathbf{E}\|_2^2}{\partial C} = 0,$$

or:

$$\sum_{k=1}^{n} 2(Ce^{Ax_k} - y_k)Cx_k e^{Ax_k} = 0$$

$$\sum_{k=1}^{n} 2(Ce^{Ax_k} - y_k)e^{Ax_k} = 0,$$

which in turn leads to the 2×2 system:

$$\begin{cases} C \sum_{k=1}^{n} x_k e^{2Ax_k} - \sum_{k=1}^{n} x_k y_k e^{Ax_k} = 0 \\ \\ C \sum_{k=1}^{n} e^{Ax_k} - \sum_{k=1}^{n} y_k e^{Ax_k} = 0 \end{cases}$$

This is not a linear system of equations any longer. It is a nonlinear system that needs to be tackled with a nonlinear solver such as Newton's method. It is subject to all the problems peculiar to such systems, in particular the lack of a general theory for the existence and uniqueness of a solution.

To avoid this difficulty, the exponential fit can be recast as a linear curve fit by the transformation:

$$Y = \ln(y), \quad X = x, \quad B = \ln(C). \tag{6.2}$$

The logarithm of the fit function $f(x) = y = Ce^{Ax}$ then becomes

$$\ln(f(x)) = \ln(y) = \ln(C) + \ln(e^{Ax}) = B + AX.$$

Thus, what is needed now is to fit the transformed set of data points:

$$\{(x_k, y_k)\} \longrightarrow \{(x_k, \ln(y_k))\} = \{(X_k, Y_k)\}$$

to the straight line $AX + B$. This amounts to minimizing the redefined error:

$$\|\tilde{\mathbf{E}}\|_2^2(A, C) = \sum_{k=1}^{n} (Ax_k + \ln C - \ln y_k)^2.$$

Obviously, once A and $B = \ln C$ have been found, the exponential fit is known from equation 6.2.

The script 6.2 exemplifies the process. A data sample is constructed from an exponential by perturbing the values of the function with random noise. Notice that the values of the coefficients produced by this code will change when the code is run repeatedly, due to the added noise being randomly generated. The fit is then plotted against the data; a sample output is shown in figure 6.2.

MATLAB Script 6.2: Exponential Fit

```
x = linspace(0,4, 60);
y = 3*exp(0.35*x);
noise = 0.5*exp(0.37*x) .* ( 2 * rand(1,length(x) ) - 1 );
ydata = y + noise;
plot( x, y, x, ydata, '*')
x2 = x .* x;
M = [ sum(x2) sum(x) ; ...
sum(x) length(x)]      % the matrix
% the new logarithmic data
ylog = log(ydata);
xy = x .* ylog;
rhs = [sum(xy) ; sum(ylog)];
coeff = M \ rhs;        % solve the system
A = coeff(1) % sample output: 0.34578
B = coeff(2);
C = exp(B) % sample output: 2.9953
yfit = C * exp(A * x)
plot(x, ydata, '*', x, yfit)
```

While this procedure does avoid the problems related to solving a system of nonlinear equations, it does introduce a difference in the weights assigned to the values in the data set; due to the nature of the logarithmic function, lower values of y are weighted more heavily. Therefore, the method has to be used with caution, or adjusted to compensate for this problem. One of the proposed ways to do so is by modifying the function to be minimized to read:

$$\|\tilde{\mathbf{E}}\|_2^2(A, C) = \sum_{k=1}^{n} y_k (Ax_k + B - \ln y_k)^2,$$

which leads to the modified system of equations:

$$
\begin{pmatrix}
\sum_{k=1}^{n} x_k^2 y_k & \sum_{k=1}^{n} x_k y_k \\
\sum_{k=1}^{n} x_k y_k & \sum_{k=1}^{n} y_k
\end{pmatrix}
\begin{pmatrix} A \\ B \end{pmatrix}
=
\begin{pmatrix}
\sum_{k=1}^{n} x_k y_k \ln y_k \\
\sum_{k=1}^{n} y_k \ln y_k
\end{pmatrix}.
$$

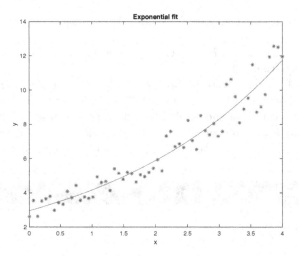

Figure 6.2: Basic exponential fit to noisy data.

This procedure extends to a number of other functions that have an inverse, for example to fits of the form $\sin(Ax+B)$. Nevertheless, more general methods are needed to deal with more fits that require more than just two parameters.

6.2   Global Polynomial Interpolation

The problem of *interpolation* can be stated as follows: one is again given a **finite** set of data points, i.e. $\{(x_k, y_k)\}$, $k = 0, 1, 2, \ldots n$. These points are considered to be **exact** values in the sense that, as opposed to curve fitting, there supposedly exists some (presumably unknown) function $f(x)$ such that $y_k = f(x_k)$. The problem is to evaluate this function, i.e. obtain $y = f(x)$, for some other input x that satisfies $x \in [x_0, x_n]$ but is not part of the data set. Note that, for this problem, it is convenient to start numbering our data points starting with $k = 0$; also, without loss of generality, we consider the data points to be ordered. The problem of finding values for x outside of the data set range $[x_0, x_n]$ is closely related and is referred to as *extrapolation*.

Most of us have used linear interpolation (i.e. interpolation using a first-degree polynomial) to find values in between two data points when working with tabulated data. Let us now take a closer look at interpolation in a more general setting. Since polynomials are well understood and easy to handle, they are natural candidates for interpolation. Our first goal is to construct a polynomial that goes through all the $(n+1)$ points $\{(x_k, y_k)\}$ in the data set. Once such polynomial is known, it can be evaluated at any point x to obtain an approximation to the function value $f(x)$.

6.2.1 Constructing the Interpolant

Thus, suppose we look for a polynomial of degree less or equal to n that passes through all the $(n+1)$ data points. This will be called the *global polynomial interpolant* based on the data set $\{(x_k, y_k)\}$. The first obvious question is whether such a polynomial exists in the first place. The following theorem provides the answer:

***Theorem* 6.2.1:** Suppose the $n+1$ points $x_0, x_1, \ldots, x_n$ are distinct, that is $x_i \neq x_j$ for $i \neq j$. Then for any arbitrary real values $y_0, y_1, \ldots, y_n$ there exists a unique polynomial $p_n(x)$ of degree less or equal to n that obeys $p_n(x_i) = y_i$ for all $i = 0, 1, 2, \ldots, n$.

A constructive proof for the existence of $p_n(x)$ can be given as follows: let us start by writing this polynomial in the standard form:

$$p_n(x) = a_0 + a_1 x + \ldots + a_n x^n.$$

Obviously, in order to define $p_n(x)$ one needs to find the coefficients $a_0, a_1, \ldots a_n$. Conversely, if the coefficients can be found, then the polynomial is well-defined through them and hence its existence is established. Notice that some of these coefficients may be zero; for example, if all the points are on a straight line, all the coefficients except a_0 and a_1 will end up being zero. To compute the coefficients, all that we need is to ask that the polynomial pass through the data points $(x_0, y_0), (x_1, y_1) \ldots (x_n, y_n)$. That means that the following equations must all be obeyed simultaneously:

$$p_n(x_0) = a_0 + a_1 x_0 + \ldots + a_n x_0^n = y_0$$
$$p_n(x_1) = a_0 + a_1 x_1 + \ldots + a_n x_1^n = y_1$$
$$\vdots$$
$$p_n(x_n) = a_0 + a_1 x_n + \ldots + a_n x_n^n = y_n.$$

These equations constitute a system of $(n+1)$ equations with $(n+1)$ unknowns $a_0, a_1, \ldots a_n$. The system can be solved easily with the MATLAB backslash command. Introducing the system quantities,

$$M = \begin{bmatrix} 1 & x_0 & x_0^2 & \cdots & x_0^n \\ 1 & x_1 & x_1^2 & \cdots & x_1^n \\ \vdots & & & & \\ 1 & x_n & x_n^2 & \cdots & x_n^n \end{bmatrix}, \quad \mathbf{a} = \begin{bmatrix} a_0 \\ a_1 \\ \vdots \\ a_n \end{bmatrix},$$

one needs to solve for $\mathbf{a}$ in:

$$M\mathbf{a} = \begin{bmatrix} y_0 \\ y_1 \\ \vdots \\ y_n \end{bmatrix}.$$

It can be shown that the determinant of the matrix M is given by

$$\det(M) = \Pi_{0 \le j < k \le n}(x_k - x_j),$$

so it is obviously not equal to zero if the points are distinct. Thus this system always has a unique solution for any arbitrary values $y_0, y_1, \ldots, y_n$.

The problem with this direct approach of constructing the polynomial is that the matrix M becomes ill-conditioned as n gets large. In fact, it turns out that matrices of this form are difficult enough to solve with finite-precision arithmetic to warrant a special name: they are known as *Vandermonde matrices*. The ill-conditioning of the system matrix means that the values of $a_0, a_1, \ldots a_n$ are subject to errors due to the finite precision of computer arithmetic.

For most of us this standard way of setting up the interpolation polynomial seems straightforward, and there appears that we have to cope with this problem or else forget about constructing this polynomial. However, there are other ways to write the same polynomial which, while not immediately obvious, are less prone to numerical error and just as straightforward to evaluate on a computer. Specifically, consider writing the polynomial $p_n(x)$ as:

$$p_n(x) = y_0 L_0(x) + y_1 L_1(x) + \ldots + y_n L_n(x),$$

where $L_0(x), L_1(x), \ldots, L_n(x)$ are themselves polynomials of degree n that have the following property:

$$L_i(x_j) = \begin{cases} 0, & i \ne j \\ 1, & i = j \end{cases}$$

Thus:

$$\begin{aligned} p_n(x_0) &= y_0 L_0(x_0) + y_1 L_1(x_0) + \ldots + y_n L_n(x_0) \\ &= y_0 \cdot 1 + y_1 \cdot 0 + \ldots + y_n \cdot 0 \\ &= y_0, \end{aligned}$$

and similarly, $p_n(x_k) = y_k$ for all $k = 0, 1, \ldots, n$.

The only remaining task is to set up the new polynomials $L_i(x)$. These will be called here the *Lagrange interpolating polynomials* for the particular data set $\{(x_k, y_k)\}$. It is relatively easy to come up with the necessary expression. Note that:

$$L_0(x) = \frac{(x - x_1)(x - x_2) \cdot \ldots \cdot (x - x_n)}{(x_0 - x_1)(x_0 - x_2) \cdot \ldots \cdot (x_0 - x_n)}$$

because this expression evaluates to 0 in $x = x_1$, $x = x_2$, ..., $x = x_n$ and to 1 for $x = x_0$. Extending to all other indices:

$$L_i(x) = \frac{(x - x_1)(x - x_2) \cdot \ldots \cdot (x - x_n)}{(x_i - x_1)(x_i - x_2) \cdot \ldots \cdot (x_i - x_n)},$$

or in shorthand notation:

$$L_i(x) = \prod_{\substack{j=0 \\ j \neq i}}^{n} \frac{(x - x_j)}{(x_i - x_j)}$$

and

$$p_n(x) = \sum_{i=0}^{n} y_i L_i(x). \tag{6.3}$$

The MATLAB function 6.3 computes the Lagrange polynomial $L_i(x)$. It uses a very useful convention for translating the mathematical notation into MATLAB code. More specifically, to go through all the data points, the length of the data vector N is used instead of the polynomial order n. Obviously $N = n + 1$, and the product $\prod_{\substack{j=0 \\ j \neq i}}^{n}$ is evaluated in a MATLAB **for** loop ranging from $j = 1$ to $j = N$ instead. Also note that, for efficiency reasons, the loop is again subdivided into two loops to avoid repeated comparison of the loop index j with the polynomial index i. Finally, the script 6.4 uses this function to plot the Lagrange polynomials of degree $n = 3$ for a set of four equidistant points. The plots generated by this script for $L_0(x)$ and $L_1(x)$ are shown in figure 6.3 and figure 6.4, respectively.

MATLAB Script 6.3: Lagrange Interpolant

```
function [Li] = lagrangeInterp( i, x, xData )
Li = 1;
% MATLAB vectors start at one
N = length(xData);

for j = 1:(i-1)
    Li = Li * ( x - xData(j) ) / ( xData(i) - xData(j) );
end
for j = (i+1):N
    Li = Li * ( x - xData(j) ) / ( xData(i) - xData(j) );
end

end
```

MATLAB Script 6.4: Script for plotting L(x)

```
clear all;
close all;
xData = 1:4; %four points, cubic polynomial
% Fine grid for plotting
xFine = linspace( 1, 4, 50);
i = 1; % in math notation, this is L_0(x)
for k = 1:length(xFine)
    L_xFine(k) = lagrangeInterp( i, xFine(k), xData );
end
% L should be zero at all data points
yData = zeros(size(xData));
% Except for point xData(i)
yData(i) = 1;
% Plot the polynomial and the data
plot( xData, yData, '*', xFine, L_xFine, 'k')
grid on
xlabel("x")
ylabel("y")
```

6.2.2   Interpolation Error

Even with infinite precision arithmetic, one cannot expect the global interpolation polynomial $p_n(x)$ to coincide with $f(x)$ unless the latter is a polynomial function itself and a sufficient number of points is used in this particular case, so that $p_n(x)$ ends up with the same degree as the original $f(x)$. On the other hand, approximation of a function by a polynomial is important for developing other numerical methods, for example for differentiation and integration. The following result, closely related to approximation of a function by Taylor series, is important for understanding how the interpolation error behaves at points that are not in the data set:

Theorem 6.2.2: Let $f(x)$ be a function that is continuous together with all its derivatives up to and including order $n+1$. Then for any point $x \in (x_0, x_n)$

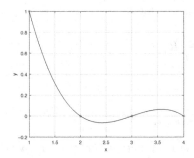

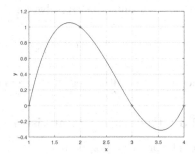

Figure 6.3: Lagrange interpolant $L_0(x)$.

Figure 6.4: Lagrange interpolant $L_1(x)$.

the error in the approximation of $f(x)$ by a polynomial of degree n constructed on the points $x_0, x_1, \ldots, x_n$ is given by:

$$f(x) - p_n(x) = \frac{f^{n+1}(\xi)}{(n+1)!} R_{n+1}(x), \tag{6.4}$$

where the point $\xi \in (x_0, x_n)$ and $R_{n+1}(x) = (x - x_0)(x - x_1) \ldots (x - x_n)$.

To prove this result, define the auxiliary function

$$E(x) = f(x) - p_n(x) - R_{n+1}(x) \frac{f(z) - p_n(z)}{R_{n+1}(z)}$$

where z is a fixed point inside the interval, distinct from the data points. It is easy to see that $E(x)$ is zero at all the data points x_k, since at these points $f(x_k) = p_n(x_k)$ and $R_{n+1}(x_k) = 0$. The value $E(z)$ is also zero because of the way this function was constructed. Thus, $E(x)$ has at least $n + 2$ zeros in the interval $[x_0, x_n]$ and by Rolle's theorem its derivative $E'(x)$ will have at least $n + 1$ zeros. Following along this line, $E''(x)$ will have at least n zeros and so on; $E^{(n+1)}(x)$ must therefore have at least one zero. Let this zero be denoted by ξ, so that we can write:

$$E^{(n+1)}(\xi) = f^{(n+1)}(\xi) - p_n^{(n+1)}(\xi) - R_{n+1}^{(n+1)}(\xi) \frac{f(z) - p_n(z)}{R_{n+1}(z)} = 0.$$

Since the polynomial $p_n(x)$ is of degree less or equal to n, its $(n + 1)$-th derivative is zero. The remainder term $R_{n+1}(x)$ is also a polynomial of degree $(n + 1)$, so its derivative is the constant $R_{n+1}^{(n+1)}(\xi) = (n + 1)!$. Thus, the fact that the $(n + 1)$-th derivative of $E(x)$ is zero reduces to:

$$f^{(n+1)}(\xi) - (n+1)! \frac{f(z) - p_n(z)}{R_{n+1}(z)} = 0.$$

Now one can immediately solve for $f(z) - p_n(z)$ to get:

$$f(z) - p_n(z) = \frac{f^{n+1}(\xi)}{(n+1)!} R_{n+1}(z),$$

which is the result stated in theorem 6.2.2.

6.2.3   Recursive Pointwise Interpolation

The interpolating polynomials that approximate a function $f(x)$ using a given set of data points can also be generated in a recursive fashion, starting from lower degrees and moving up to higher degrees. This can be done either point-wise, to evaluate the value of the function at a given point with increasing accuracy, or globally, to obtain the polynomial expression. The pointwise approach is discussed in this section.

To illustrate the idea, focus on the case of a data set consisting of three points, $\{(x_0, y_0), (x_1, y_1), (x_2, y_2)\}$ and consider that $y_i = f(x_i)$ for some unknown function f. For $i = 0, 1, 2$, denote by $P_i(x)$ the degree-zero (i.e. constant) polynomial that takes the value $P_i(x) = y_i$ for all values of x. Although these three polynomials might constitute poor candidates for an interpolating polynomial in general, either one of them could be used to interpolate the function at a point ξ. That is, one can use the zeroth-order approximations $f(\xi) = P_i(\xi) = y_i$, especially for points ξ close to the point x_i. With this in mind, notice that there are also three approximations with polynomials of degree one which can be built from the same data set. Using subscript indices to indicate the two data points involved, the three polynomials, written in Lagrange form, are:

$$P_{0,1}(x) = y_0 \frac{x - x_1}{x_0 - x_1} + y_1 \frac{x - x_0}{x_1 - x_0} = P_0(x) \frac{x - x_1}{x_0 - x_1} + P_1(x) \frac{x - x_0}{x_1 - x_0}$$

$$P_{1,2}(x) = y_1 \frac{x - x_2}{x_1 - x_2} + y_2 \frac{x - x_1}{x_2 - x_1} = P_1(x) \frac{x - x_2}{x_1 - x_2} + P_2(x) \frac{x - x_1}{x_2 - x_1}$$

$$P_{0,2}(x) = y_0 \frac{x - x_2}{x_0 - x_2} + y_2 \frac{x - x_0}{x_2 - x_0} = P_0(x) \frac{x - x_2}{x_0 - x_2} + P_2(x) \frac{x - x_0}{x_2 - x_0}.$$

One can then rewrite the above polynomials in the form:

$$P_{0,1}(x) = y_0 \frac{x - x_1}{x_0 - x_1} - y_1 \frac{x - x_0}{x_0 - x_1} = \frac{(x - x_1)P_0(x) - (x - x_0)P_1(x)}{x_0 - x_1}$$

and similarly:

$$P_{1,2}(x) = \frac{(x - x_2)P_1(x) - (x - x_1)P_2(x)}{x_1 - x_2},$$

$$P_{0,2}(x) = \frac{(x - x_2)P_0(x) - (x - x_0)P_2(x)}{x_0 - x_2}.$$

This form of the linear interpolants highlights a procedure that can be extended to obtain, in a straightforward manner, an immediately higher-degree interpolant once two interpolants are already available. As an example, the only degree-two polynomial based on this data set can be written in the form:

$$P_{0,1,2}(x) = \frac{(x - x_2)P_{0,1}(x) - (x - x_0)P_{1,2}(x)}{x_0 - x_2}, \tag{6.5}$$

the proof of which is left as an exercise. A similar formula is obtained by combining a different choice of first degree polynomials, for example using $P_{0,1}$ and $P_{0,2}$ to obtain the same $P_{0,1,2}$. Notice also that this latter polynomial coincides with the global interpolant that uses all the data points, which was previously denoted by $p_2(x)$.

The extension of this approach to obtaining pointwise values of one-degree-higher interpolants from lower-degree values that are already available is due to Neville. To put the discussion in a proper framework, consider a data set $\{(x_i, y_i)\}_{i=0,1,\ldots,n}$, where the values y_i are thought of as the values of a (possibly unknown) function, that is $y_i = f(x_i)$ for some f. The data point locations are also supposed to be distinct: $x_i \neq x_j$ if $i \neq j$. Choose a subset $i_1, i_2, \ldots, i_k$ of k distinct integers and let $P_{i_1, i_2, \ldots, i_k}(x)$ be the polynomial that interpolates the function f at the points in the subset, that is

$$P_{i_1 i_2 \ldots i_k}(x_{i_j}) = y_{i_j}$$

for all the points $x_{i_1}, x_{i_2}, \ldots, x_{i_k}$. All the indices for the points in the subset are naturally supposed to satisfy $0 \leq i_k \leq n$, with $k \leq n + 1$. Notice that, with this notation, the global interpolant for this data set can be written: $p_n(x) = P_{0,1,\ldots,n}(x)$

The following result generalizes the previous three-point example:

Theorem 6.2.3: Let x_i, x_j be two x values in the data set (x_k, y_k), $k = 0, 1, \ldots, n$. Then the global interpolant for the full data set can be written as:

$$p_n(x) = \frac{(x - x_j)P_{0,1,\ldots,j-1,j+1,\ldots,n}(x) - (x - x_i)P_{0,1,\ldots,i-1,i+1,\ldots,n}(x)}{x_i - x_j}.$$

That is, $p_n(x)$ as given above has the property that $p_n(x_k) = y_k$ for all $k = 0, 1, \ldots, n$.

While this theorem does provide a way to obtain general expressions of (higher-degree) interpolants as a general function of the variable x, it is most commonly used to produce pointwise sequences of values that approximate the unknown function at a fixed location. To see how this works, it is useful to simplify the notation. The notation with multiple indices (i.e. $P_{0,1,2}$) is very general and powerful, but it can get cumbersome to use when the index set is large. Following [5], let $Q_{i,j}(x)$ for $0 \leq j \leq i$ denote the **degree-j** interpolating polynomial on the $(j + 1)$ positions $x_{i-j}, x_{i-j+1}, \ldots, x_{i-1}, x_i$.

That is, the connection between this latter and the more general notation is therefore:

$$Q_{i,j}(x) = P_{i-j,\,i-j+1,\,...,\,i-1,\,i}(x),$$

and a straightforward application of the theorem above to the subset $\{(x_k, y_k)\}_{k=0,1,...,i}$ leads to:

$$Q_{i,j}(x) = \frac{(x - x_{i-j})Q_{i,j-1}(x) - (x - x_i)Q_{i-1,j-1}(x)}{x_i - x_{i-j}}, \quad j = 1, 2, \ldots, i.$$

This relationship only needs to be applied for $j \geq 1$, because the case $j = 0$ is readily available, $Q_{i,0}(x) = y_i$. It can be used to build a table of approximate values for the interpolated function $f(x)$ once a point x is chosen in the interpolation range, as shown in the following example.

Example. Using the data for $y = f(x)$ shown in table 6.2, which is obtained with $f(x) = e^{-x}$, the goal is to estimate $f(x = 0.5)$.

TABLE 6.2: Initial data for $y = f(x) = e^{-x}$.

k	0	1	2	3	4
x_k	0.0	0.2	0.3	0.6	0.7
y_k	1.00000	0.81873	0.74082	0.54881	0.49659

Then:

$$Q_{0,0}(x) = y_0 = 1.0$$
$$Q_{1,0}(x) = y_1 = 0.81873,$$

etc, where $Q_{k,0}(x) = y_k$ is the zeroth-order approximation for the function $f(x)$ at $x = 1.5$ that makes exclusive use of the value of the function at x_k. Without extra computations, someone will naturally suppose that the best such approximation is $Q_{3,0}(x)$, since x_3 is closest to the evaluation point $x = 0.5$. Suppose now that one wants to compute approximations, based on the same data, of successively higher degree for $f(0.5)$. The linear approximations are:

$$Q_{1,1}(0.5) = \frac{(0.5 - 0.0)Q_{1,0} - (0.5 - 0.2)Q_{0,0}}{0.2 - 0.0}$$
$$= \frac{0.5(0.81873) - 0.3(1.0)}{0.2}$$
$$\approx 0.54682.$$

Similarly:

$$Q_{2,1}(0.5) = \frac{(0.5 - 0.2)(0.74082) - (0.5 - 0.3)(0.81873)}{0.3 - 0.2}$$

$$\approx 0.585$$

$$Q_{3,1}(0.5) \approx 0.61281$$

$$Q_{4,1}(0.5) \approx 0.60103.$$

Since $x = 0.5$ falls between $x_2 = 0.3$ and $x_3 = 0.6$, one may expect the best degree-one approximation to be $Q_{3,1}(0.5)$.

In a similar manner the approximations using quadratic polynomials are given by:

$$Q_{2,2}(1.5) = \frac{(0.5 - 0.0)(0.585) - (0.5 - 0.3)(0.54682)}{0.3 - 0.0}$$

$$\approx 0.61045$$

$$Q_{3,2}(1.5) \approx 0.60586$$

$$Q_{4,2}(1.5) \approx 0.60692.$$

The higher degree approximations can be generated in a similar manner, and the entries conveniently arranged as shown in table 6.3. If more accuracy is desired, another data point can be added to the table, as needed. In fact, such a table is usually not computed from the very beginning. Instead, one proceeds by adding data points (lines in the table) then augments a lower-degree approximation by calculating the extra column made possible by the newly added points. Thus, this method can be easily used to generate successive approximations at a given, specified location.

It is worth emphasizing again that all the entries in the table represent different approximations to the value of $f(0.5)$. Each column is for polynomial fits of a fixed degree, and the different rows represent different data subsets on which the approximations are based. The only value in the last column uses all points in the original data set (and hence is supposedly the most accurate). Some entries in the table represent values obtained by extrapolation.

||

6.2.4   Newton's Divided Differences

The divided difference method is another way which can be used to generate recursively interpolating polynomials of increasing degree. Unlike the method in the previous section, it is not commonly used for pointwise values but rather for constructing the expression of the global interpolant. Suppose that one is given again the $(n + 1)$ points $(x_0, y_0), \ldots, (x_n, y_n)$ and the values of y are thought of as a function $y = f(x)$. Define, for each $i = 0, 1, \ldots, n$ the

TABLE 6.3: Recursive approximation for $e^{(-0.5)} = 0.6065306597$.

x_j	$Q_{j,0}$	$Q_{j,1}$	$Q_{j,2}$	$Q_{j,3}$	$Q_{j,4}$
0.0	1.00000				
0.2	0.81873	0.54682			
0.3	0.74082	0.58500	0.61045		
0.6	0.54881	0.61281	0.60586	0.606625	
0.7	0.49659	0.60103	0.60692	0.606497	0.606534

zeroth divided differences of the function f at x_i as being just the values of the function at that respective point,

$$f[x_i] = f(x_i).$$

Higher ordered divided differences are defined recursively, using the immediately lower order. For each $i = 0, 1, \ldots, n - 2, n - 1$, the **first** divided difference relative to x_i and x_{i+1} is defined in terms of the zeroth-order divided differences:

$$f[x_i, x_{i+1}] = \frac{f[x_{i+1}] - f[x_i]}{x_{i+1} - x_i}.$$

After the $(k - 1)$st divided differences

$$f[x_i, x_{i+1}, \ldots, x_{i+k-1}]$$

and

$$f[x_{i+1}, x_{i+2}, \ldots, x_{i+k-1}, x_{i+k}]$$

have been determined, the k-th divided difference relative to $x_i, x_{i+1}, \ldots, x_{i+k}$ is:

$$f[x_i, x_{i+1}, \ldots, x_{i+k}] = \frac{f[x_{i+1}, \ldots, x_{i+k}] - f[x_i, \ldots, f_{i+k-1}]}{x_{i+k} - x_i}.$$

Notice that every time the order is increased, the number of divided differences that can be constructed out of a data set is reduced by one. The process can therefore proceed only until the single n-th order divided difference $f[x_0, \ldots, x_n]$, which uses all the data points, is obtained.

With this notation, the counterpart of theorem 6.2.3 is:

Theorem 6.2.4: The global interpolant for the data set $\{(x_k, y_k)\}$, $k = 0, 1, \ldots, n$ can be written in the Newton divided-difference form:

$$p_n(x) = f[x_0] + f[x_0, x_1](x - x_0) + f[x_0, x_1, x_2](x - x_0)(x - x_1) + \ldots$$
$$+ f[x_0, x_1, \ldots, x_n](x - x_0)(x - x_1) \cdot \ldots \cdot (x - x_{n-1})$$
$$= \sum_{k=0}^{n} f[x_0, x_1, \ldots, x_k] \prod_{j=0}^{k-1} (x - x_j).$$

The polynomial $p_n(x)$ shown above satisfies $p_n(x_k) = y_k$ for all $k = 0, 1, \ldots, n$.

This result can be proved by induction on n and is left as an exercise for those who have been exposed to this proof technique.

The generation of divided differences can be arranged neatly in a table similar to a Pascal's triangle (usually oriented horizontally), as in the following example.

Example. Using the data for $y = f(x)$ given in table 6.2, Newton's divided-difference formula yields the results in table 6.4. The interpolation polynomial can then be written:

$$p_4(x) = 1 - 0.90635x + 0.42417x(x - 0.2) - 0.1275x(x - 0.2)(x - 0.3)$$
$$+ 0.03048x(x - 0.2)(x - 0.3)(x - 0.6).$$

||

TABLE 6.4: Newton divided differences for data in Table 6.2.

i	x_i	$f[x_i]$	$f[x_{i-1}, x_i]$	$f[x_{i-2}, x_{i-1}, x_i]$	$\ldots$
0	0.0	1.00000			
			-0.90635		
1	0.2	0.81873		0.42417	
			-0.77910		and so on
2	0.3	0.74082		0.34767	$\longrightarrow$
			-0.64003		
3	0.6	0.54881		0.29458	
			-0.52220		
4	0.7	0.49659			

It is important to note that, when used recursively, the Newton divided-difference formula has what may be thought of as a built-in directionality mechanism. The form shown in theorem 6.2.4 is oftentimes referred to as the *forward divided-difference formula*. One can start with the only value $f(x_0)$ as a zeroth-order interpolant, and recursively add points to obtain higher-order approximations. During this process, the last point in the pool is only used when computing the last member in the sum for $p_n(x)$, while the first point x_0 appears in all the terms in the sum. Nevertheless, the x-points in the data set do not necessarily have to be in increasing order. In fact, it is often the case that points are added repeatedly around the x-location of interest until a satisfactory approximation is obtained. The ordering of the points does have an impact, however, on the propagation of round-off errors that may occur following small perturbations in the data.

Several alternative forms for Newton's divided-difference interpolation formula are available. Depending on the context, some of these forms are easier to use than others. The Newton *backward divided-difference formula* is obtained by merely reordering the points, leading to:

$$p_n(x) = f[x_n] + f[x_{n-1}, x_n](x - x_n) + \ldots$$

$$= \sum_{k=0}^{n} f[x_{n-k}, x_{n-k+1}, \ldots, x_n] \prod_{j=0}^{k-1} (x - x_{n-j}).$$

If the x-locations in the data set are equispaced, $x_k = x_0 + kh$, for $k = 0, 1, \ldots, n$, where h is the spacing between any two successive data points, these formulae can be rearranged further. Define the *forward difference* of the function f at the point x_k, Δf_k, to be:

$$\Delta f_k = f(x_{k+1}) - f(x_k) = y_{k+1} - y_k, \quad k = 0, 1, \ldots, (n-1),$$

with its powers obtained naturally as follows:

$$\Delta^0 f_k = f(x_k), \quad \Delta^1 f_k = \Delta f_k,$$

$$\Delta^2 f_k = \Delta(\Delta f_k) = \Delta f_{k+1} - \Delta f_k,$$

$$\Delta^m f_k = \Delta(\Delta^{m-1} f_k), \quad m = 2, 3, \ldots.$$

Introducing a non-dimensional coordinate of the point x where the polynomial is to be evaluated and a generalization of the binomial coefficient notation to non-integral values,

$$s = \frac{x - x_0}{h}, \quad \begin{pmatrix} s \\ k \end{pmatrix} = \frac{s(s-1)\ldots(s-k+1)}{k!}, \quad \begin{pmatrix} s \\ 0 \end{pmatrix} = 1$$

the forward divided-difference formula can be written

$$p_n(s) = \sum_{k=0}^{n} \begin{pmatrix} s \\ k \end{pmatrix} \Delta^k f_0.$$

This is reffered to as the *Newton forward difference* form. In a similar way, the *Newton backward difference* form can be obtained with the use of the *backward difference* of the function at the point x_k, $\nabla f_k = y_k - y_{k-1}, k = 1, 2, \ldots, n$. In this case, it can be verified by direct calculation that the backward formula becomes:

$$p_n(t) = \sum_{k=0}^{n} (-1)^k \begin{pmatrix} -t \\ k \end{pmatrix} \nabla^k f_n, \tag{6.6}$$

where now the non-dimensional coordinate is defined as $t = (x - x_n)/h$ and the binomial coefficient is extended to negative non-integral values,

$$\begin{pmatrix} -t \\ k \end{pmatrix} = \frac{-t(-t-1)\ldots(-t-k+1)}{k!} = (-1)^k \frac{t(t+1)\ldots(t+k-1)}{k!}.$$

6.2.5 Hermite interpolation

In a more general setting, one may have access not only to the values of the function at the data points, but also to the values of its derivative up to some order M. That is, for *Hermite interpolation* the data set is of the form $F_0 \cup F_1 \cdots \cup F_M$. Here $F_0 = \{(x_k, y_k)\}_{k=0,1,\dots,n}$ is the usual data set which contains the function values at the sample points, while $F_m = \{(x_k, y_k^{(m)})\}_{k=0,1,\dots,n}$ provides the values of the m-th derivative of the function f which is to be approximated at the same points, $f^{(m)}(x_k) = y_k^{(m)}$. Since there are now $(n+1)(M+1)$ conditions available, a polynomial of degree $(n+1)(M+1) - 1$ can be constructed. Consider for example the case $n = M = 2$, so that the interpolating polynomial will have the form:

$$p_3(x) = a_0 + a_1 x + a_2 x^2 + a_3 x^3$$

and its derivative:

$$p_3'(x) = a_1 + 2a_2 x + 3a_3 x^2.$$

Imposing the interpolation conditions $p_3(x_0) = y_0$, $p_3'(x_0) = y_0'$ etc. leads then to the following system of equations

$$\begin{pmatrix} 1 & x_0 & x_0^2 & x_0^3 \\ 1 & x_1 & x_1^2 & x_1^3 \\ 0 & 1 & 2x_0 & 3x_0^2 \\ 0 & 1 & 2x_1 & 3x_1^2 \end{pmatrix} \begin{bmatrix} a_0 \\ a_1 \\ a_2 \\ a_3 \end{bmatrix} = \begin{bmatrix} y_0 \\ y_1 \\ y_0' \\ y_1' \end{bmatrix}.$$

It can be shown that this system has a unique solution when the points x_k are distinct. However, approached in this manner, Hermite interpolation may very quickly become prohibitive, as the cost of solving the system by Gauss elimination scales roughly with $(n+1)(M+1)^3$. In order to avoid this, the interpolating polynomial may be expressed in a way that uses the Lagrange form; only the case $M = 1$ is presented here in detail. Under this scenario the polynomial degree is $(M+1)(n+1) - 1 = 2n + 1$ and the polynomial is to be written as

$$p_{2n+1}(x) = \sum_{k=0}^{n} y_k H_{0k}(x) + \sum_{k=0}^{n} y_k' H_{1k}(x)$$

in terms of some basis polynomials $H_{0k}(x)$ and $H_{1k}(x)$, all of the same degree $2n+1$. To satisfy the interpolation conditions, these polynomials must satisfy:

$$H_{0k}(x_\ell) = 0, \; k \neq \ell; \quad H_{0k}(x_k) = 1; \quad H_{0k}'(x_\ell) = 0, \forall \ell;$$

$$H_{1k}'(x_\ell) = 0, \; k \neq \ell; \quad H_{1k}'(x_k) = 1; \quad H_{1k}(x_\ell) = 0, \forall \ell.$$

One can then notice that these basis polynomials can be written in terms of the Lagrange interpolants $L_k(x)$, which are polynomials of order n, in the form:

$$H_{0k}(x) = (Ax + B)L_k^2(x), \quad H_{1k}(x) = (Cx + D)L_k^2(x),$$

where A, B, C and D are coefficients that may depend on the point x_k and are to be determined. This is convenient because it immediately enforces the conditions $H_{0k}(x_\ell) = H_{1k}(x_\ell) = 0$ for $k \neq \ell$ due to the fact that $L_k(x_\ell) = 0$ in that case. Therefore, there are only four other conditions to be satisfied, which allows us to determine the four coefficients A, B, C and D:

$$H_{0k}(x_k) = (Ax_k + B)L_k^2(x_k) = (Ax_k + B) = 1,$$

$$H'_{0k}(x_k) = AL_k^2(x_k) + (Ax_k + B)\left(2L'_k(x_k)L_k(x_k)\right) = A + 2L'_k(x_k) = 0,$$

$$H_{1k}(x_k) = (Cx_k + D)L_k^2(x_k) = (Cx_k + D) = 0,$$

$$H'_{1k}(x_k) = CL_k^2(x_k) + (Cx_k + D)\left(2L'_k(x_k)L_k(x_k)\right) = C = 1,$$

where the last equation in the sequence made use of the third. Solving for the coefficients yields:

$$A = -2L'_k(x_k); \quad B = 1 + 2x_k L'_k(x_k); \quad C = 1; \quad D = -x_k,$$

and hence the interpolation polynomial can be written:

$$
\begin{aligned}
p_{2n+1}(x) &= \sum_{k=0}^{n} y_k(1 - 2L'_k(x_k)(x - x_k))L_k^2(x) \\
&+ \sum_{k=0}^{n} y'_k(x - x_k)L_k^2(x).
\end{aligned}
\tag{6.7}
$$

A first MATLAB implementation of this formula, which you are asked to improve upon in the exercises, is given in script 6.5. It reuses the code that computes the Lagrange interpolant, shown in script 6.3. For the example function chosen here, $f(x) = x \sin(x)$, the exact value at all the sample points is zero (the computed MATLAB values are slightly different due to machine precision). The Hermite polynomial does a very good job at interpolating the function, as can be seen in figure 6.5. On the other hand, the polynomial $p_4(x)$ computed from equation 6.3 using only the Lagrange polynomials $L_k(x)$, which does not account for the derivatives, would in this case be identically zero.

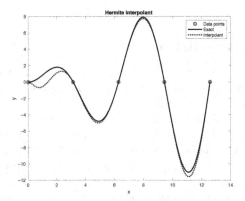

Figure 6.5: Hermite interpolation for $f(x) = x \sin(x)$.

MATLAB Script 6.5: Hermite Interpolation

```
xData = linspace(0, 4*pi, 5);
f = @(x) x .* sin(x);
df = @(x) sin(x) + x .* cos(x);
yData = f(xData);
dy = df(xData);
xPlot = linspace(0, 4*pi, 100);

% evaluate the Hermite polynomial
for j = 1 : length(xPlot)
  x = xPlot(j);
  yInt(j) = hermite(x, xData, yData, dy );
end
plot ( xData, yData, 'o', xPlot, f(xPlot), '-k', xPlot, yInt,
    ':k',...
    'LineWidth',2 )

function [s] = hermite (x, xData, yData, ypData)
  s = 0; % need to compute a double sume
  for k = 1:length(xData)
    sq = lagrangeInterp(k,x,xData)^2;
    xk = xData(k);
    s = s + yData(k) * ( 1 - 2*dL(k, xData)*(x-xk ) ) * sq;
    s = s + ypData(k) * (x - xk ) * sq;
  end
end


function [val] = dL(k, xData)
  val = 0;
  for j = 1:(k-1)
    val = val + 1 / ( xData(k) - xData(j) );
  end
  for j = (k+1):length(xData)
    val = val + 1 / ( xData(k) - xData(j) );
  end
end
end
```

6.3   Spline Interpolation

While it may be argued that relatively high-degree polynomials are needed for accurate approximations, these have some serious disadvantages. In particular, as the chapter exercises highlight, unless due care is taken in choosing the interpolation nodes the interpolants may exhibit oscillatory behavior. Nevertheless, in many situations there is little freedom in placing the nodes; in such cases an alternative approach is needed. A sensible solution is then to divide the interval $[x_0, x_n]$ into a collection of subintervals and construct a different, relatively low-order, approximating polynomial on each interval. This will then be a piecewise polynomial approximation with low-order polynomials which are less likely to display large oscillations.

The simplest form of such a construction is a piecewise linear polynomial. A graph of such an approximation can be seen in figure 6.6. This is commonly used for tabulated data, and many of us have encountered it in one form or another. The disadvantage of this approximation is that the first derivative is not continuous across subinterval boundaries; the interpolating function is not necessarily differentiable at the endpoints of the subintervals. This is a major drawback, since access to an approximation of the first derivative is required in many numerical approaches to solving differential equations.

The most common piecewise interpolation approximation uses cubic polynomials between pairs of nodes. This is known as a *cubic spline* approximation. A general cubic involves four coefficients, which translates into the fact that continuity of both the first and the second derivatives can be imposed. However, the approximation doesn't require that the derivatives of the cubic spline will necessarily agree with the derivatives of the interpolated function (which may not be known in fact) even at the interpolation nodes $x_0, x_1, \ldots, x_n$.

6.3.1   A simple example

Consider a data set containing three data points, (x_0, y_0), (x_1, y_1) and (x_2, y_2), which delineate two patches on the interval $[x_0, x_2]$. The goal is to build two cubic polynomials, $S_0(x)$ and $S_1(x)$ such that the following piecewise function:

$$S(x) = \begin{cases} S_0(x), x \in [x_0, x_1) \\ S_1(x), x \in [x_1, x_2] \end{cases}$$

interpolates the data while also being continuous and ensuring continuity of the first and second derivatives at the common point x_1. The two cubic polynomials could be written in the form:

$$S_j(x) = A_j + B_j x + C_j x^2 + D_j x^3, \quad j \in \{0, 1\},$$

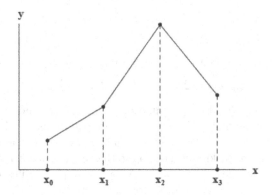

Figure 6.6: Piecewise linear function.

but it turns out that this form leads to more complicated calculations. The equivalent form

$$S_j(x) = a_j + b_j(x - x_j) + c_j(x - x_j)^2 + d_j(x - x_j)^3$$

simplifies the work and will be considered here. To obtain the values of the eight coefficients that are required one needs to use eight equations. The first three such equations are immediately obvious: in order to interpolate the data, the coefficients need to satisfy the following:

$$S(x_0) = S_0(x_0) = y_0; \; S(x_1) = S_1(x_1) = y_1; \; S(x_2) = S_1(x_2) = y_2.$$

Because:

$$S_j(x = x_j) = a_j + b_j(x_j - x_j) + c_j(x_j - x_j)^2 + d_j(x_j - x_j)^3 = a_j,$$

the three conditions can be written as:

$$a_0 = y_0; \; a_1 = y_1; \; a_1 + b_1 h_1 + c_1 h_1^2 + d_1 h_1^3 = y_2, \tag{6.8}$$

where the notation $h_1 = x_2 - x_1$ was introduced for convenience. Next, continuity of the interpolant $S(x)$ at the inner grid point x_1 requires that $S_0(x_1) = S_1(x_1)$, leading to a fourth equation among the coefficients:

$$S_0(x_1) = a_0 + b_0(x_1 - x_0) + c_0(x_1 - x_0)^2 + d_0(x_1 - x_0)^3 = S_1(x_1) = a_1. \tag{6.9}$$

This can again be written in a more convenient form by introducing $h_0 = x_1 - x_0$ and noticing that $a_1 = y_1$,

$$S_0(x_1) = a_0 + b_0 h_0 + c_0 h_0^2 + d_0 h_0^3 = S_1(x_1) = y_1, \tag{6.10}$$

similar to the last equation in 6.8. Taking into consideration that

$$S_j'(x) = b_j + 2c_j(x - x_j) + 3d_j(x - x_j)^2; \; S_j''(x) = 2c_j + 6d_j(x - x_j), \tag{6.11}$$

continuity of the first derivative at x_1 leads to:

$$S'_0(x_1) = b_0 + 2c_0h_0 + 3d_0h_0^2 = b_1 = S'_1(x_1), \tag{6.12}$$

while continuity of the second derivative requires that

$$S''_0(x_1) = 2c_0 + 6d_0h_0 = 2c_1 = S''_1(x_1). \tag{6.13}$$

There are thus a total of six equations that enforce the requirements set forth for our interpolant $S(x)$ so far. Two more equations are therefore needed to find the eight coefficients. These two equations are somewhat arbitrary. One can choose to use extra knowledge about the function that is to be interpolated, if such knowledge is available. As an example, values of the first or second derivatives at the two endpoints x_0 and x_2 could be used. Most often, if no such knowledge is available, the second derivative at the two endpoints is simply required to be zero. This leads to the following two extra conditions:

$$S''(x_0) = S''_0(x_0) = 2c_0 = 0; \; S''(x_1) = S''_1(x_1) = 2c_1 + 6d_1h_1 = 0.$$

With these, the system to be solved becomes:

$$
\begin{pmatrix}
1 & 0 & 0 & 0 & 0 & 0 & 0 & 0 \\
0 & 0 & 0 & 0 & 1 & 0 & 0 & 0 \\
0 & 0 & 0 & 0 & 1 & h_1 & h_1^2 & h_1^3 \\
1 & h_0 & h_0^2 & h_0^3 & 0 & 0 & 0 & 0 \\
0 & 1 & 2h_0 & 3h_0^2 & 0 & -1 & 0 & 0 \\
0 & 0 & 2 & 6h_0 & 0 & 0 & -2 & 0 \\
0 & 0 & 2 & 0 & 0 & 0 & 0 & 0 \\
0 & 0 & 0 & 0 & 0 & 0 & 2 & 6h_1
\end{pmatrix}
\begin{bmatrix}
a_0 \\ b_0 \\ c_0 \\ d_0 \\ a_1 \\ b_1 \\ c_1 \\ d_1
\end{bmatrix}
=
\begin{bmatrix}
y_0 \\ y_1 \\ y_2 \\ y_1 \\ 0 \\ 0 \\ 0 \\ 0
\end{bmatrix}. \tag{6.14}
$$

It can easily be seen that some of the equations in this system can be simplified further, for example by dividing throughout by a factor of two in the last two equations. What is not immediately obvious is whether the system always has a unique solution. The answer to this question can be obtained easier, once some of the unknowns in this system are eliminated. It will then be reduced to only the equations involving the coefficients c_j. This also opens the door to a systematic generalization and is explored in the next section.

6.3.2 General Formulation

Consider again a data set $\{(x_k, y_k)\}, k = 0, 1, \ldots, n$ that represents known, tabulated data about the values of a function f, that is $f(x_k) = y_k$ for all k. Although we'll see that we can incorporate some extra information about the function in this context, in the most general case we consider that nothing else is known about the function other than these pointwise values. The interpolation points are supposed to be such that $a = x_0 < x_1 < \ldots < x_n = b$ and will be referred to, collectively, as a grid. A cubic spline interpolant $S : [a, b] \to \mathbb{R}$ built on this data set is a function of one variable that satisfies the following conditions:

(S1) It is a piecewise function, that is:

$$S(x) = \begin{cases} S_0(x), x \in [a = x_0, x_1) \\ S_1(x), x \in [x_1, x_2) \\ \vdots \\ S_{n-2}(x), x \in [x_{n-2}, x_{n-1}) \\ S_{n-1}(x), x \in [x_{n-1}, x_n = b] \end{cases}$$

(S2) $S(x_j) = y_j$ for $j = 0, 1, \ldots, n$. This states that the piecewise function $S(x)$ recovers the values in the data set, i.e. it is an interpolant for that particular data set.

(S3) $S_{j+1}(x_{j+1}) = S_j(x_{j+1})$ for $j = 0, 1, \ldots, (n-2)$. This enforces the continuity of $S(x)$ at the interior grid points.

(S4) $S'_{j+1}(x_{j+1}) = S'_j(x_{j+1})$ for $j = 0, 1, \ldots, (n-2)$. This enforces the continuity of the first derivative of the interpolant $S(x)$ at all interior grid points.

(S5) $S''_{j_1} = S''_j(x_{j+1})$ for $j = 0, 1, \ldots, (n-2)$. This enforces the continuity of the second derivative at all interior grid points.

It is now time to do a little bookkeeping and see if the conditions stated above are enough for the construction of the interpolant. Using the same form as for the simple example in the previous section, the cubic polynomial is written as:

$$S_j(x) = a_j + b_j(x - x_j) + c_j(x - x_j)^2 + d_j(x - x_j)^3,$$

for $j = 0, 1, 2, \ldots, n-1$. It is obvious that full knowledge of $S(x)$ requires knowledge of the $4n$ coefficients, $\{a_j, b_j, c_j, d_j\}$, $j = 0, 1, \ldots, (n-1)$, on the n patches that make up the grid. Condition (S2) translates into $(n+1)$ equations, while conditions (S3) through (S5) translate into $(n-1)$ equations each, for a total of $(4n-2)$ equations; this should make clear the need for two extra conditions. Thus we need to consider one more set of constraints:

(S6) Two additional equations, needed to uniquely determine the spline coefficients. Any additional knowledge about the function f which is to be interpolated can be used here. These extra conditions usually involve derivatives of the function f and take one of the following two forms:

(S6.1) $S''(x_0) = S''(x_n) = 0$ (known as natural or free boundary conditions) or:

(S6.2) $S'(x_0) = f'(x_0)$ and $S'(x_n) = f'(x_n)$ (known as clamped boundary conditions)

Unless the derivative of $f(x)$ at the boundaries is available, the natural boundary conditions (S6.1) are the default choice. The development below uses this scenario and the treatment of clamped boundary conditions is left as an exercise.

The form of the piecewise polynomial makes its evaluation at the grid points easy, as if follows immediately that

$$S_j(x_j) = a_j = y_j,$$

because of (S2). Therefore, the continuity condition (S3) can be written in the form:

$$
\begin{aligned}
a_{j+1} &= S_{j+1}(x_{j+1}) \\
&= S_j(x_{j+1}) \\
&= a_j + b_j(x_{j+1} - x_j) + c_j(x_{j+1} - x_j)^2 + d_j(x_{j+1} - x_j)^3,
\end{aligned}
$$

for each index j in the range $j = 0, 1, \ldots, n - 2$.

As we have seen in the previous subsection, the quantities $x_{j+1} - x_j$ will appear repeatedly. Let's again denote these values by:

$$h_j = x_{j+1} - x_j, \quad j = 0, 1, \ldots, n - 1.$$

Also, although there is no polynomial $S_n(x)$, it is still convenient to introduce a fictitious coefficient a_n defined by $a_n = y_n$ in order to unify notation. With this, it can be seen that the equation

$$a_{j+1} = a_j + b_j h_j + c_j h_j^2 + d_j h_j^3 \tag{6.15}$$

holds for each index j in the range $j = 0, 1, 2, \ldots, n - 1$.

To obtain the equations that stem from (S4), notice that

$$S_j'(x) = b_j + 2c_j(x - x_j) + 3d_j(x - x_j)^2$$

and therefore $S_j'(x_j) = b_j$ for each $j = 0, 1, \ldots, n-1$. Therefore $S_{j+1}'(x_j+1) = S_j'(x_j + 1)$ leads to

$$b_{j+1} = b_j + 2c_j h_j + 3d_j h_j^2 \tag{6.16}$$

for any $j = 0, 1, \ldots, n - 2$. Although this will not be pursued here, it is sometimes convenient, for example when the two conditions (S6.2) are used instead of the more common (S6.1), to define $b_n = S'(x_n)$ in a similar manner to a_n and extend this equation up to $j = n - 1$.

Proceeding further, the second derivative of $S_j(x)$ at the point x_j can be seen to be $S_j''(x_j) = 2c_j$, in view of equation (6.11). Applying condition (S5) leads then to the following relationship upon division by a factor of two on both sides:

$$c_{j+1} = c_j + 3d_j h_j, \tag{6.17}$$

for $j = 0, 1, 2, \ldots, n-2$. For natural boundary conditions it is helpful to again define $c_n = S''(x_n)/2 = 0$ and extend this equation to $j = 0, 1, 2, \ldots, n - 1$.

The coefficient d_j can now be obtained from equation (6.17) in terms of c_j and c_{j+1}. Doing so and substituting back into equations (6.15) and (6.16), respectively, produces two new equations:

$$a_{j+1} = a_j + b_j h_j + \frac{h_j^2}{3}(2c_j + c_{j+1}) \tag{6.18}$$

and

$$b_{j+1} = b_j + h_j(c_j + c_{j+1}) \tag{6.19}$$

where $j = 0, 1, \ldots, n-1$ for the first equation and $j = 0, 1, \ldots, n-2$ for the second. Finally, solving for b_j in equation (6.18) produces:

$$b_j = \frac{1}{h_j}(a_{j+1} - a_j) - \frac{h_j}{3}(2c_j + c_{j+1}), \tag{6.20}$$

a relationship that holds for $j = 0, 1, \ldots, (n-1)$. This can be rewritten, shifting the index j by one unit through $j = J - 1$, as:

$$b_{J-1} = \frac{1}{h_{J-1}}(a_J - a_{J-1}) - \frac{h_{J-1}}{3}(2c_{J-1} + c_J), \tag{6.21}$$

where now $J = 1, 2, \ldots, n$. Since the index is just a dummy variable that only serves to indicate the range of validity of a relationship, this last equation can be written just as well as:

$$b_{j-1} = \frac{1}{h_{j-1}}(a_j - a_{j-1}) - \frac{h_{j-1}}{3}(2c_{j-1} + c_j), \tag{6.22}$$

with $j = 1, 2, \ldots, n$.

Performing a similar shift of the index j in equation (6.19), substituting the expressions of b_j and b_{j-1} from equations (6.20) and (6.22) and rearranging terms leads to:

$$\begin{aligned} h_{j-1}c_{j-1} + 2(h_{j-1} + h_j)c_j + h_j c_{j+1} &= \tfrac{3}{h_j}(a_{j+1} - a_j) \\ &\quad - \tfrac{3}{h_{j-1}}(a_j - a_{j-1}) \end{aligned} \tag{6.23}$$

which is now valid only for $j = 1, 2, \ldots, n-1$ due to the more restricted range of equation (6.20).

This $(n-1)$ relationships form a system of equations for the coefficients c_j, because the quantities on the right-hand side are known (recall that $a_j = y_j$ are the function values in the data set and $h_j = x_{j+1} - x_j$ represents the grid spacing between the nodes). Also notice that the first and last of these equations involve the values of the second derivative at the end points through c_0 and c_n; these unknowns can either be eliminated from the system (leading to a tridiagonal system of equations) or added as two additional equations that enforce their values. The latter approach is considered here; the matrix of the system obtained this way is strictly diagonally dominant and hence it can be shown that it has a unique solution as long as the grid points are distinct. Therefore, one can solve this system of equations for the unknown coefficients c_j, $j = 1, 2, \ldots, n-1$. Once the unknown coefficients c_j are found, b_j and d_j can be obtained, for example from equations (6.20) and (6.17), respectively.

In the case of the clamped boundary, the above procedure has to be modified to impose $S'(x_0) = f'(x_0)$, $S'(x_n) = f'(x_n)$. Using the expression for $S'_j(x)$, it is left as an exercise to show that the first and last equations in the system (6.23) need to be modified to read:

$$2h_0c_0 + h_0c_1 = \frac{3}{h_0}(a_1 - a_0) - 3f'(x_0)$$

$$\text{and} \tag{6.24}$$

$$h_{n-1}c_{n-1} + 2h_{n-1}c_n = 3f'(x_n) - \frac{3}{h_{n-1}}(a_n - a_{n-1}).$$

6.3.3   Setting up the spline Interpolant

For the case of natural boundary conditions, i.e. $c_0 = c_n = 0$, the coefficients c_k, $k = 0, 1, \ldots, n$ must be obtained as the solution of the linear system

$$\begin{bmatrix} & & \\ & A & \\ & & \end{bmatrix} \begin{bmatrix} c_0 \\ c_1 \\ c_2 \\ \vdots \\ c_n \end{bmatrix} = \begin{bmatrix} \\ RHS \\ \\ \end{bmatrix},$$

where, given the fact that the quantities h_j are positive, it is easy to see that the matrix A is strictly diagonally dominant and given by:

$$A = \begin{bmatrix} 1 & 0 & 0 & 0 & \cdots & 0 \\ h_0 & 2(h_0 + h_1) & h_1 & 0 & \cdots & 0 \\ 0 & h_1 & 2(h_1 + h_2) & h_2 & \cdots & 0 \\ & & \ddots & & & \\ & & & \ddots & & \\ 0 & \cdots & 0 & h_{n-2} & 2(h_{n-2} + h_{n-1}) & h_{n-1} \\ 0 & 0 & 0 & 0 & \cdots & 1 \end{bmatrix}$$

and

$$RHS = \begin{bmatrix} 0 \\ \frac{3}{h_1}(a_2 - a_1) - \frac{3}{h_0}(a_1 - a_0) \\ \frac{3}{h_2}(a_3 - a_2) - \frac{3}{h_1}(a_2 - a_1) \\ \vdots \\ 0 \end{bmatrix}.$$

To write a function that computes these coefficients, remember that in MATLAB matrix indices start from one instead of zero. Suppose the data is supplied as (xdata, ydata); then a function that sets up the system and solves for $c(1), c(2), \ldots, c(N)$ (which correspond in our notation to $c_0, c_1, \ldots c_n$), where $N = \texttt{length(xdata)} = n + 1$ can be written as follows:

Matlab Function 6.1: Finding Spline Coefficients

```
function [c] = splineCs(xdata, ydata)
N = length(xdata);
A = zeros (N, N);
A(1,1) = 1;
A(N,N) = 1;
h = diff(xdata);
for i = 2:(N-1)
    A(i,i) = 2*(h(i) + h(i-1));
    A(i, i-1) = h(i-1);
    A(i, i+1) = h(i);
    rhs(i) = (3/h(i))*(ydata(i+1) - ydata(i)) - ...
        (3/h(i-1))*(ydata(i) - ydata(i-1))
end
rhs(1) = 0;
rhs(N) = 0;
% now the matrix and RHS are set up,
% solve for the coefficients c
c = A \ rhs';   % rhs' because rhs is by default a row vector
end
```

In order to compute the remaining coefficients of the spline's piecewise cubic polynomials we now use:

(a) Equation 6.20:

$$b_j = \frac{1}{h_j}(a_{j+1} - a_j) - \frac{h_j}{3}(2c_j + c_{j+1}),$$

$j = 0, 1, \ldots, n - 1$.

(b) Equation 6.17:

$$d_j = \frac{c_{j+1} - c_j}{3h_j},$$

$j = 0, 1, 2, \ldots, n - 1$.

The code for finding b_j and d_j can be readily added to MATLAB function 6.1, which could be modified to return the three vectors in that case. Alternately, a different function that takes the coefficients c_j already computed as input and returns the coefficients b_j and d_j could be contemplated.

6.4   Approximation with Trigonometric Functions

One of the most important tasks that physicists and engineers regularly face is *signal analysis*: identifying the frequency content of, and possibly reproducing, oscillatory signals. This is routinely done nowadays with devices whose functioning is based on ideas due to the French mathematician and physicist J. Fourier. In the simplest setting, consider a function f defined on the whole real axis which is periodic with period P; that is, $f(x) = f(x + P)$ for all x. It is common, but not necessary, to take $P = 2\pi$. We will also consider here that the function is continuous everywhere on any interval of length P, except possibly for a finite number of discontinuities, themselves of finite size. The *Fourier series* generated by f, denoted S, is a series of trigonometric functions that approximates f and is written in the form of

$$f(x) \sim S(x) = \frac{a_0}{2} + \sum_{n=1}^{\infty} \left[a_n \cos\left(\frac{2\pi nx}{P}\right) + b_n \sin\left(\frac{2\pi nx}{P}\right) \right], \qquad (6.25)$$

where the coefficients of the sine and cosine functions in the series are given by:

$$\begin{aligned} a_n &= \frac{2}{P} \int_{\alpha}^{\alpha+P} f(x) \cos\left(\frac{2\pi nx}{P}\right) dx, \\ b_n &= \frac{2}{P} \int_{\alpha}^{\alpha+P} f(x) \sin\left(\frac{2\pi nx}{P}\right) dx, \end{aligned} \qquad (6.26)$$

and the lower limit of the integral α can safely be taken to be any convenient value. The symbol $\sim$ reminds us that the series is a formal expansion – this formal series of functions on the right may diverge in many cases, and one needs to clearly understand the extent to which it approximates f.

a_n and b_n are called the *Fourier coefficients*. The integral forms of a_n and b_n are related to the trigonometric integrals

$$\int_{\alpha}^{\alpha+P} \sin\left(\frac{2\pi mx}{P}\right) \cos\left(\frac{2\pi nx}{P}\right) = 0, \qquad (6.27)$$

$$\int_{\alpha}^{\alpha+P} \cos\left(\frac{2\pi mx}{P}\right) \cos\left(\frac{2\pi nx}{P}\right) = \begin{cases} \dfrac{P}{2} & m = n, \\ 0 & m \neq n, \end{cases} \qquad (6.28)$$

Here m and n are integers. Note that we can replace the cosine functions in (6.28) by sine functions and reach the same conclusion. The trigonometric functions in a Fourier series form the *Fourier basis*. Equations (6.27) and (6.28) indicate that Fourier basis is an orthogonal basis.

A common choice, which emphasizes once again the relationship to the common trigonometric functions, is $\alpha = -\pi$ and $P = 2\pi$, leading to:

$$a_n = \frac{1}{\pi} \int_{-\pi}^{\pi} f(x) \cos(nx)\, dx$$

and similarly for b_n. Notice that the trigonometric functions that are involved in the series oscillate faster with increasing n, as $\cos(2\pi nx/P)$ exhibits n cycles within the period. When the variable x is time and is measured in seconds, the reciprocal of the period, $1/P$, is known as the *fundamental frequency* and its unit of measure is the *hertz*. The coefficients a_n and b_n, the amplitudes of the sines and cosines that add up to the original function, contain information about the frequency content of the original function at each frequency that is an integer multiple of the fundamental. These frequencies are known as the *harmonics*; the amplitude of the n-th harmonic is given by $R_n^2 = a_n^2 + b_n^2$.

Suppose the sequence of partial sums of the Fourier series (6.4) is $\{S_N\}_{N=1}^{\infty}$, where

$$S_N(x) = \frac{a_0}{2} + \sum_{n=1}^{N-1} \left[a_n \cos\left(\frac{2\pi nx}{P}\right) + b_n \sin\left(\frac{2\pi nx}{P}\right) \right].$$

Let S_N, $N = 1, 2, 3, \ldots$ be defined on an interval I (which may be chosen to be either closed or open). Suppose f is a continuous function on I; it is known that the sequence $\{S_N\}_{N=1}^{\infty}$ converges (pointwise) to S and $S(x) = f(x)$, for $x \in I$. Mathematically, we write

$$\lim_{N\to\infty} S_N(x) = S(x) = f(x)$$

for all $x \in I$.

One may consider a weaker condition for f, for example assume f is a piecewise continuous function in I. Then S_N converges to $\bar{f}$ pointwise on I, where

$$\bar{f}(x) := \frac{1}{2}\left(f(x_+) + f(x_-) \right),$$

is the average of the one-sided limits at x. Clearly at points where f is continuous, $\bar{f} = f$. The Fourier series for a function can thus be seen as a type of global approximation since it converges in an average sense at discontinuities.

For many practical problems, f is not periodic. It is desirable to extend the method of Fourier series to include nonperiodic functions. Without going into the detailed derivation, we write the *Fourier integral* for a general nonperiodic function f as

$$f(x) = \frac{1}{\pi} \int_0^{\infty} [A(w)\cos wx + B(w)\sin wx]\, dw, \tag{6.29}$$

where

$$A(w) = \int_{-\infty}^{\infty} f(v)\cos wv\, dv, \quad B(w) = \int_{-\infty}^{\infty} f(v)\sin wv\, dv. \tag{6.30}$$

Plugging (6.30) into (6.29) and using the trigonometric identity

$$\cos wx \cos wv + \sin wx \sin wv = \cos w(x - v)$$

and the double integral

$$0 = \frac{i}{2\pi} \int_{-\infty}^{\infty} \left[\int_{-\infty}^{\infty} f(v) \sin w(x - v) dv \right] dw, \qquad (6.31)$$

where $i = \sqrt{-1}$, one can relate the Fourier integral, (6.29) and (6.30), to the *Fourier Transform* pair

$$f(x) = \frac{1}{2\pi} \int_{-\infty}^{\infty} F(w)e^{ixw} dw,$$

$$F(w) = \int_{-\infty}^{\infty} f(x)e^{-ivw} dv. \qquad (6.32)$$

The Fourier Transform has a wide range of applications in physical sciences, mathematical sciences, and engineering. It is an important tool for signal and image processing, including image analysis, image filtering, image reconstruction and image compression. The discrete Fourier transform (DFT) approximates the Fourier Transform by a finite sum over the first N frequencies (Fourier modes), where N is a finite integer. A fast Fourier Transform (FFT) is a very efficient algorithm that computes the DFT, or its inverse (IDFT) [9]. The FFT algorithms, available in MATLAB, are not discussed here in detail, but examples of their use are provided.

6.5   MATLAB® Built-in Functions

6.5.1   Piecewise Interpolation

While it is useful to know how interpolation routines work, and to be able to write codes that perform this task, MATLAB already offers a number of built-in functions that perform curve fitting and interpolation. The first one is `interp1`, which does piecewise polynomial interpolation; the degree of the piecewise polynomial can be adjusted, using optional parameters, from a constant up to a cubic, in which case a cubic spline will be obtained. The simplest call to the function will return a piecewise linear interpolant; a piecewise constant, staircase-like approximation can be obtained with the `nearest`, `next` or `previous` options. The script 6.6 shows a couple of these options, with the result depicted in figure 6.7. There are several other options that can be used for `interp1`, including `spline`, which will create a cubic spline, and `extrapolation` if some of the evaluation points fall outside of the data x-range.

MATLAB Script 6.6: Using interp1.

```
clear all;
xData = [ 0.1, 0.3, 0.5, 0.9, 1.5];
yData = [ 2.1, 3.5, 3.7, 2.2, 3.5];
xEval = 0.1:0.01:1.5;
yInt = interp1(xData, yData, xEval,'nearest');
subplot(1,2,1);
plot(xEval, yInt, 'k', xData, yData, '*')
xlabel('x')
ylabel('y')
title('Piecewise constant interpolant')
yInt = interp1(xData, yData, xEval);
subplot(1,2,2);
plot(xEval, yInt, 'k', xData, yData, '*')
xlabel('x')
ylabel('y')
title('Piecewise linear interpolant')
```

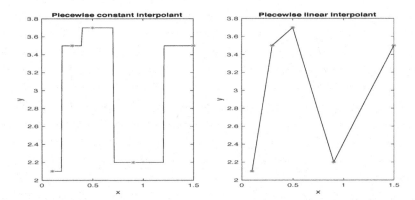

Figure 6.7: Piecewise constant and piecewise linear interpolants created with `interp1`.

The function **griddedInterpolant** conveniently wraps the same type of interface for one- and multi-dimensional interpolation around calls to **interp1**. Its use in one dimension to create the spline interpolant shown in figure 6.8 for the same data is depicted in script 6.8.

MATLAB Script 6.7: Using griddedInterpolant.

```
xData = [ 0.1, 0.3, 0.5, 0.9, 1.5];
yData = [ 2.1, 3.5, 3.7, 2.2, 3.5];
F = griddedInterpolant(xData, yData, 'spline');

xEval = 0.1:0.01:1.5;
yEval = F(xEval);
plot(xData,yData,'k*')
hold on
plot(xEval,yEval,'k-')
legend('Data Points','Interpolated Values')
```

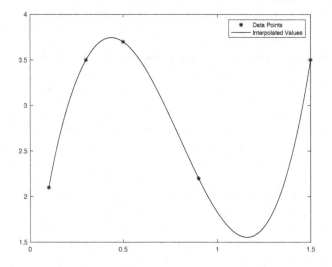

Figure 6.8: Spline created by `interp1` via call to `griddedInterpolant`.

For convenience, the `spline` MATLAB function can also be used:

>>yEval = spline(xData, yData, xEval)

which saves some typing but otherwise is the same as `interp1` with the `'spline'` option.

6.5.2   Polynomial fits to data

The MATLAB built-in function `polyfit` can used to generate the $(n + 1)$ coefficients of the global nth degree polynomial passing through the $(n + 1)$ data points $\{x_k, y_k\}_{k=0,1,\dots,n}$. If the data is stored in two vectors, labeled `xdata` and `ydata` below, then the corresponding call to the function is:

>> pcoeff = polyfit(xdata, ydata, n)

Notice that `polyfit` also accepts as a third argument the order of the polynomial to be created from the data; the value of the third parameter must satisfy `n < length(xdata)`. The coefficients of the global interpolation polynomial $p_n(x)$ will be returned in the case `n = length(xdata) - 1`, while for smaller values of n the function returns the coefficients of the least-squares fit of corresponding degree to the data. For example,

```
>> pcoeff = polyfit(xdata, ydata, 1)
```

will return the two coefficients of the straight line fit for the data points. Once polynomial coefficients are available, the polynomial can be evaluated with the built-in function `polyval`, which uses these coefficients together with the a vector containing the points where the polynomial is to be evaluated and produces the corresponding vector of values at these points. The script 6.8 shows different ways to use these functions and produces figure 6.9. As a somewhat related problem, MATLAB also offers a way to build a piecewise polynomial with specified (possibly different) orders by providing distinct sets of coefficients on different patches through the functions `mkpp` and `ppval`.

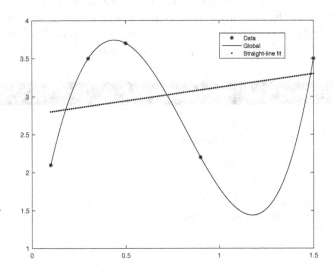

Figure 6.9: Using `polyfit` to fit data.

MATLAB Script 6.8: Using griddedInterpolant.

```
xData = [ 0.1, 0.3, 0.5, 0.9, 1.5];
yData = [ 2.1, 3.5, 3.7, 2.2, 3.5];
n = length(xData);
xEval = 0.1:0.01:1.5;
% calculate the global interpolant
pcoeff = polyfit(xData, yData, n)
yEval = polyval(pcoeff, xEval);
% straight-line fit
pcoeff = polyfit(xData, yData, 1)
yEval1 = polyval(pcoeff, xEval);
% quadratic fit
pcoeff = polyfit(xData, yData, 2)
yEval2 = polyval(pcoeff, xEval);
% plot
plot(xData, yData, 'k*', xEval, yEval, 'k-', ...
    xEval, yEval1, 'k.')
legend('Data', 'Global', 'Straight-line fit')
```

A way to use the Fourier transform to approximate functions via the built-in MATLAB command `interpft` is shown in script 6.9 with the corresponding figure 6.10. Note that the input data points `xData` must be equispaced when using `interpft`. Also notice that the number of computed values needs to be adjusted by recalculating the spacing of the interpolation points `xPlot` as shown.

MATLAB Script 6.9: Using interpft.

```
clear all;
xData = linspace(0, 2*pi, 20);
yData = sin(xData) .* sin(2*xData);
N = 100;
yInt = interpft(yData, N);
dx = xData(2) - xData(1);
dxPlot = dx*length(xData)/N;
xPlot = 0:dxPlot:2*pi;
yPlot = yInt(1:length(xPlot));
plot( xPlot, yPlot, 'k-', xData, yData, 'ko')
title('Approximation with Fourier series')
xlabel('x')
ylabel('y')
```

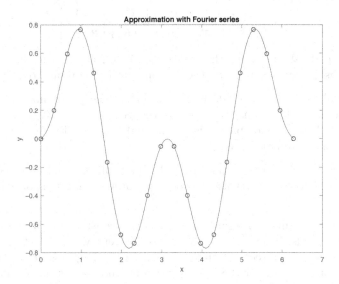

Figure 6.10: Using `interpft` to approximate a periodic function.

6.6   Exercises

Ex. 1 Consider the problem of finding a quadratic fit to a data set (x_i, y_i), $i = 1, 2, \ldots, n$, which was introduced in section 6.1.1. Evaluate the partial derivatives in equation 6.1 to obtain the three equations that need to be satisfied by the coefficients A, B and C.

Ex. 2 Write a MATLAB script that implements the quadratic fit in the problem above. Run the script on the data in the following table:

x	0	1	2	3	4
y	0.695	−1.475	−1.275	0.882	4.765

Print your values for the three coefficients A, B and C, then plot $f(x)$ vs x where `x = 0:0.1:4` along with the original data points in the same figure.

Ex. 3 Write up, then solve, the system for the coefficients of the polynomial of degree two going through the points $(1, 3)$, $(2, 7)$ and $(-1, 1)$. Then write up the expansion of the same polynomial in terms of the Lagrange interpolants $L_k(x)$. Do all the calculations required to bring the Lagrange form to the standard form and confirm that you obtain the same polynomial.

Ex. 4 Write another script file to find an exponential fit of the form $f(x) = Ce^{Ax}$, to the data

x	0.0	0.1	0.2	0.3	0.4	0.5
y	1.388	1.647	1.951	2.633	3.321	3.977

The code should print the values for A and C, then again plot $f(x)$ versus x, where x = 0:0.01:0.5 along with the original data points.

Ex. 5 Calculate the standard-form coefficients of the polynomial that interpolates the function $f(x) = e^x$ at the points $x_k = -0.2 + k \cdot 0.1$, for $k = 0, 1, 2, 3, 4$. Compare this polynomial with the Taylor series expansion of the function around $x = 0$.

Ex. 6 Show, by direct calculation, that the interpolant based on three points can be written in terms of two-point interpolants as shown in equation (6.5).

Ex. 7 Prove theorem 6.2.3 regarding the recursive construction of interpolating polynomials. There are at least two different ways to get the result: one can use a direct calculation based on the Lagrange form, as in the previous exercise, or the fact that two polynomials of degree at most n that take the same value in $n + 1$ distinct points are identical.

Ex. 8 Use the data set in the following table:

k	0	1	2	3
x_k	0.1	0.3	0.5	0.7
y_k	1.1	1.25	1.17	1.31

to approximate the value of the function $y = f(x)$ at the point $x = 0.45$. To do this, construct the table of values $Q_{ij}(x)$ for the recursive point-wise evaluation of the interpolation polynomials. Do all computations by hand, rounding off to three decimal places.

Ex. 9 Approximate the value of the same function $y = f(x)$ tabulated in the previous problem at the same point $x = 0.45$. This time do the computation by constructing the complete table of Newton's divided differences. Using these results, explicitly write the interpolation polynomial in divided differences form. Finally, extract both a linear and a quadratic approximation for $f(0.45)$ from this final result. Again do this calculation by hand by rounding off to three decimal places.

Ex. 10 Based on the Newton's backward divided-difference formula, obtain by direct calculation the non-dimensional form in equation (6.6). You may start by writing out the first two terms to get a feeling for the calculations that are involved, then move on to address the general term in the sum.

Ex. 11 For the function $f(x) = 0.5x^3 - 2x^2 + 3x + 1$, use Newton's divided differences to find the interpolating polynomials of degrees two, three and four using the points $x = -2, -1, 0, 1, 2$. Plot the three resulting approximations against the original function in the range $[-2, 2]$ using one hundred equally-spaced points.

Ex. 12 Write a MATLAB function that calculates and prints out the full table of divided differences for a given data set. The data is provided as two vectors, xData and yData.

Ex. 13 Consider the data in the following table:

k	0	1	2	3
x_k	0.0	1.761062	3.522123	5.283185
y_k	1.0	-0.1891196	-0.9284676	0.5403023

The values of y in this table have been obtained as $y_k = \cos(x_k)$. The goal of this exercise it to create three different spline interpolants obtained by using these data points, and compare them with the original function.

(a) Write a code to compute the usual spline interpolant $^1S(x)$ that uses the natural boundary conditions. Plot this interpolant versus the original function $\cos(x)$ using 100 data points equally spaced between x_0 and x_3. Also indicate the four tabulated data points on the plot using a special marker such as $*$ or $\circ$ for example.

(b) For most people the comparison at point (a) above doesn't look very good. This is mostly due to the fact that the natural boundary conditions do not match the behavior of the $\cos(x)$ function at the end points. The task now is to modify the spline code to account for the correct second derivatives at the endpoints. Create a new interpolant $^2S(x)$ that satisfies the following conditions:

$$^2S''(x_0) = -\cos(x_0); \quad ^2S''(x_3) = -\cos(x_3).$$

You can do this by modifying the system of equations for the spline coefficients accordingly. Plot again the newly-obtained interpolant versus the original function in the same manner as above.

(c) Use the MATLAB built-in function **spline** to compute and plot a new interpolant $^3S(x)$ that uses the same data points but imposes clamped boundary conditions at the endpoints.

Ex. 14 Make any necessary changes to the MATLAB script 6.5, including the function **hermite**, to abide by the standard MATLAB behavior in that it would accept the full vector xPlot as input argument instead of only a scalar value, and consequently return the full vector yInt as output.

Ex. 15 Prove to your own satisfaction that the MATLAB function dL in script 6.5 returns the value of the derivative of the Lagrange interpolant $L'_k(x_k)$. That is, show that

$$L'_k(x_k) = \sum_{j=0, j \neq k}^{n} \frac{1}{x_k - x_j}.$$

Ex. 16 Values $y_k = f(x_k)$ are provided in the table below. It is known that the function f is a polynomial of degree $n \leq 5$. Find the actual degree of the polynomial as well as the coefficients of the lowest and highest powers of x.

k	0	1	2	3	4	5	6
x_k	1.0	1.5	2.0	2.5	3.0	3.5	4.0
y_k	6.20	5.6375	4.20	2.1875	0.20	−0.8625	0.2

Ex. 17 Consider the function defined for all real values x by

$$f(x) = \frac{1}{1 + 14x^2}.$$

This is an example of a *Runge function*, a classical example that showcases the problems that high-order global interpolation leads to.

(a) Compute the interpolating polynomial for $f(x)$ using a data set $\{(x_k, y_k)\}$, where x_k are equally spaced nodes in $[-1, 1]$ with

$$-1 = x_0 < x_1 < \ldots < x_n = 1.$$

and y_k is the value of the function at each node (i.e. $y_k = f(x_k)$). Use $n = 9$ (i.e. 10 nodes).

(b) Repeat the calculation for part (a) but this time use nodes that are not equispaced. Instead, use the Chebyshev nodes $\{x_k\}$ given by

$$x_k = -\cos(\pi * k/n), \quad k = 0, 1, \ldots, n$$

again with $n = 9$.

Remember that in our mathematical notation the points are numbered starting from zero. To create a MATLAB vector you will need to shift all indices by one, including the index k in the formula for the Chebyshev nodes. For both cases, plot the data points and the resulting interpolant against the original function on a set of one hundred equally spaced nodes in the range $[-1, 1]$. Repeat the same calculation for $n = 20$.

Ex. 18 Consider again the same Runge function as in the previous problem. Compute the cubic spline interpolant for $f(x)$ using the same data set $\{(x_k, y_k)\}$ as in part (a) of the previous problem. Again, plot it against the data points and the original function.

Ex. 19 One more time, construct an approximation for the same Runge function, this time using Fourier series with both $n = 9$ and $n = 20$. Note that in this case the function will be considered periodic. Plot the interpolant and compare it with the original function.

Chapter 7

Numerical Differentiation

Techniques for evaluating derivatives and integrals form the backbone of many tools used nowadays to analyze physical systems. Derivatives describe how certain quantities change with regards to space, time, or both. For example, given a set of values that represent the position of a particle in time, the first and second derivatives represent the velocity and acceleration of that particle, respectively. Conservation laws that govern physical systems are therefore expressed naturally in terms of these rates of change and lead to equations involving derivatives. These are known as differential equations, may contain either ordinary or partial derivatives, and in most cases of interest require numerical solutions based on the methods to be discussed in the following chapters.

7.1 Basic Derivative Formulae

Before starting this section it is useful to emphasize that, as is generally the case in a numerical framework, the functions for which derivative approximations are sought are supposed to only be available through their discrete, pointwise values. A first step in exploring how to obtain numerical derivative formulae is to start from the definition of the first derivative of a function f from calculus:

$$\frac{df(t)}{dx} = \lim_{h \to 0} \frac{f(x+h) - f(x)}{h}. \tag{7.1}$$

In the case of numerical evaluations one can hope that, for $\triangle x$ relatively small, the quantity

$$\frac{f(x + \triangle x) - f(x)}{\triangle x}$$

is an good approximation for the value of the derivative $f'(x)$. Because the quantity h in equation 7.1, or equivalently $\triangle x$ in the ratio above, can be negative, it is useful to already introduce some nomenclature at this stage. The point where an approximation for the derivative of a function is sought will be called the *base point* for that particular derivative approximation, while the set of all points involved in the approximation is collectively referred to as the *stencil*. For the formula above the base point is the (rather generic) point

x and the stencil is the set of points $\{x, x + \triangle x\}$. Assuming that $\triangle x > 0$, the stencil involves only one point located to the right of the base point on the real line, in which case the formula is known as a *forward* approximation to the derivative. Alternately, for $\triangle x < 0$, one obtains a *backward*. Obviously, in such numerical approximations, one needs to use non-zero, albeit small, increments $\triangle x$. These increments will interchangeably be denoted by $\triangle x = h$ in the numerical formulae to be obtained below.

While formulae such as the one above may be justified, they provide no immediate information about the magnitude of the error involved. There is, therefore, no immediately available guarantee for their accuracy. To quantify and possibly control the error associated with this type of approximation, which will obviously depend not only on the function involved and the location of the approximation but also on $\triangle x$, one can make use of Taylor series expressions. For example, if instead of the function values $f(x + \triangle x)$ and $f(x)$ one uses the two points on either side of the base point, thus letting $\{x - \triangle x, x, x + \triangle x\}$ be the stencil, and again letting $h = \triangle x > 0$ for convenience, the function values at the stencil points can be expressed in terms of the values at the base point as:

$$f(x + h) = f(x) + hf'(x) + \frac{h^2}{2!}f''(x) + \frac{h^3}{3!}f'''(x) + \cdots$$

$$f(x - h) = f(x) - hf'(x) + \frac{h^2}{2!}f''(x) - \frac{h^3}{3!}f'''(x) + \cdots$$

Subtracting these two expressions, one obtains:

$$f(x + h) - f(x - h) = 2hf'(x) + \frac{2h^3}{3!}f'''(x) + \text{H.O.T},$$

and thus:

$$\frac{df}{dx} = f'(x) = \frac{f(x + h) - f(x - h)}{2h} + \frac{h^2}{3!}f'''(x) + \text{H.O.T.} \qquad (7.2)$$

Therefore, another approximation for the fist derivative, known as a *central* approximation, can be obtained:

$$\frac{df}{dx} \approx \frac{f(x + h) - f(x - h)}{2h},$$

for which the error has the form:

$$\text{error} = \frac{h^2}{3!}f'''(x) + \text{higher-order terms in } h.$$

Note that the largest term in the error is proportional to h^2. The smaller h is, the smaller the error is expected to become, and this decrease is proportional to the second power of h. It is common to state that

$$\frac{f(x + h) - f(x - h)}{2h}$$

is an approximation of second order for the first derivative of $f(x)$ at the point x.

It can similarly be proven that the more straightforward formulae:

$$\frac{f(x+h) - f(x)}{h} \quad \text{and} \quad \frac{f(x) - f(x-h)}{h},$$

the forward and backward approximations we already saw, are accurate only to first order in h.

Other formulae can be obtained by using the values of f at even more neighboring points. The next sections will discuss the need for, and a systematic way to derive, such derivative approximation formulae. For example,

$$\frac{df}{dx} \approx \frac{-f(x+2h) + 8f(x+h) - 8f(x-h) + f(x-2h)}{12h}$$

is accurate to order four in h; as seen before, one way to conveniently state this fact is by saying that the error is $\mathcal{O}(h^4)$.

Approximating higher derivatives works in a similar fashion. For example,

$$f''(x) = \frac{f(x+h) - 2f(x) + f(x-h)}{h^2} + \mathcal{O}(h^2).$$

The following tables list several of the more often encountered formulae. The centered formulae make use of a symmetric set of stencil points around the base point, while the forward and backward formulae use a right- and leftt-biased set of points, respectively.

TABLE 7.1: Second-order centered difference formulae.

Quantity	Approximation
$f'(x)$	$\frac{f(x+h) - f(x-h)}{2h}$
$f''(x)$	$\frac{f(x+h) - 2f(x) + f(x-h)}{h^2}$
$f'''(x)$	$\frac{f(x+2h) - 2f(x+h) + 2f(x-h) - f(x-2h)}{2h^3}$
$f''''(x)$	$\frac{f(x+2h) - 4f(x+h) + 6f(x) - 4f(x-h) + f(x-2h)}{h^4}$

TABLE 7.2: Second-order forward and backward difference formulae.

Quantity	Approximation	Type
$f'(x)$	$\frac{-3f(x)+4f(x+h)-f(x+2h)}{2h}$	forward
$f'(x)$	$\frac{3f(x)-4f(x-h)+f(x-2h)}{2h}$	backward
$f''(x)$	$\frac{2f(x)-5f(x+h)+4f(x+2h)-f(x+3h)}{h^2}$	forward
$f''(x)$	$\frac{2f(x)-5f(x-h)+4f(x-2h)-f(x-3h)}{h^2}$	backward

TABLE 7.3: Fourth-order centered difference formulae.

Quantity	Approximation
$f'(x)$	$\frac{-f(x+2h)+8f(x+h)-8f(x-h)+f(x-2h)}{12h}$
$f''(x)$	$\frac{-f(x+2h)+16f(x+h)-30f(x)+16f(x-h)-f(x-2h)}{12h^2}$
$f'''(x)$	$\frac{-f(x+3h)+8f(x+2h)-13f(x+h)+13f(x-h)-8f(x-2h)+f(x-3h)}{8h^3}$
$f''''(x)$	$\frac{-f(x+3h)+12f(x+2h)-39f(x+h)+56f(x)-39f(x-h)+12f(x-2h)-f(x+3h)}{6h^4}$

7.2 Derivative Formulae Using Taylor Series

Let us start exploring the use of formulae like those derived above when approximating the derivatives at a collection of points, which will be known as the *grid*, a common name for a discretization of the range of the parameter an unknown function depends on. The simplest scenario is when the points in the grid are equally spaced. When working on a computer, the number of points in the grid is necessarily finite; there will always be a leftmost, as well as a rightmost, point in our set. This means, in turn, that one needs more than one derivative formula in order to keep the approximation at the same level of accuracy throughout the whole grid. This is known as the *closure problem*, which can be easily understood if one considers the grid points spanning an interval $[a, b]$,

$$a = x_0 < x_1 < x_2 < \ldots < x_n = b,$$

where $x_{j+1} - x_j = h$ for all $j = 0, 1, \ldots, (n-1)$. It is then easy to notice that the centered formula for the first derivative,

$$\frac{df}{dx} \approx \frac{f(x+h) - f(x-h)}{2h}$$

can only be used at the interior grid points $x_1, x_2, \ldots, x_{n-1}$. At the endpoints x_0 and x_n one can of course use the forward and backward formulae, respectively, but when doing so one order of accuracy is lost, the truncation error going from $\mathcal{O}(h^2)$ to $\mathcal{O}(h)$. It is therefore natural to ask the question: How can one keep $\mathcal{O}(h^2)$ accuracy throughout the whole grid?

The computational cost of a derivative approximation resides mainly in evaluating the values of the function. For all the first derivative approximations we have seen so far, only two function evaluations are necessary; nevertheless, for this price, the central approximation produces a more accurate result. It is to be expected, however, that using extra points (hence extra function evaluations and thus more computational work) in a one-sided formula will lead to higher accuracy. For example, suppose we use $f(x), f(x+h)$, and $f(x+2h)$ to get an approximation for $f'(x)$; this would be a one-sided, right-biased formula that could be used at the first grid point $x = x_0$. To investigate such a possibility, a straightforward approach is to use Taylor series for the function $f(x)$ around the point where the derivative approximation is sought.

Thus, the Taylor series approximations for $f(x)$ lead to the following relationships:

$$f(x+h) = f(x) + \frac{h}{1!}f'(x) + \frac{h^2}{2!}f''(x) + \frac{h^3}{3!}f'''(x) + \text{H.O.T.}$$

$$f(x+2h) = f(x) + \frac{2h}{1!}f'(x) + \frac{(2h)^2}{2!}f''(x) + \frac{(2h)^3}{3!}f'''(x) + \text{H.O.T.}$$

Since all the previous formulae use a linear combination of function values to approximate the derivative, we can begin by looking for such a linear combination of the form:

$$f'(x) \approx af(x) + bf(x+h) + cf(x+2h).$$

Making use of the Taylor series above, this linear combination becomes:

$$
\begin{aligned}
af(x) + bf(x+h) + cf(x+2h) &= (a+b+c)f(x) + (b+2c)hf'(x) \\
&+ (b+4c)\tfrac{h^2}{2!}f''(x) + (b+8c)\tfrac{h^3}{3!}f'''(x) + \text{H.O.T.} \approx f'(x).
\end{aligned}
\tag{7.3}
$$

Since one needs the linear combination on the left-hand side to be the best possible approximation to the first derivative, it is obvious that one must ask that the following relationships hold:

(1) $a + b + c = 0$ (no $f(x)$ in the approximation)

(2) $(b + 2c)h = 1$ (recover $f'(x)$)

(3) $(b + 4c)\dfrac{h^2}{2!} = 0$ (highest possible accuracy)

The first equation is necessary, because there should be no $f(x)$ term in the linear combination that is supposed to approximate $f'(x)$. The second equation recovers $f'(x)$ on the right-hand side, which is the required term. At this point, there are three coefficients, i.e. $\{a, b, c\}$ and only two equations; so there is room for one more relationship. It therefore makes sense to set to zero the coefficient that has the lowest power of h on the right-hand side, that is, the coefficient of $f''(x)$; otherwise, this will become the leading error term. On the other hand, if this coefficient is set to zero, the leading error will come from the third derivative term; thus we will gain one order of accuracy. This is an important strategy that is worth adopting every time when looking for new derivative formulae and leads to the third equation above. The system of equations obtained from (1), (2), and (3) can then be solved for a, b, and c, leading to

$$a = \frac{-3}{2h}, \quad b = \frac{4}{2h}, \quad c = \frac{-1}{2h}.$$

The leading error term will then be $(b+8c)\tfrac{h^3}{3!}f'''(x) = \tfrac{-4h^2}{3!}f'''(x)$. Note that the coefficients involved in such a linear combination, in this case a, b and c, always depend on the grid spacing h themselves.

Let us see how this same procedure can be used to obtain a higher derivative approximation, using the same stencil, again for the leftmost stencil point. The approximation will be of the same form,

$$f''(x) \approx af(x) + bf(x+h) + cf(x+2h),$$

where of course the coefficients a, b and c will be different from the preceding case. Taking into consideration the expression in equation 7.3, the coefficients now have to obey the following equations:

(1') $a + b + c = 0$ (no $f(x)$ in the approximation)

(2') $(b + 2c)h = 0$ (no $f'(x)$ in the approximation)

(3') $(b + 4c)\dfrac{h^2}{2!} = 1$ (recover $f''(x)$)

Solving for the coefficients leads this time to:

$$a = \frac{1}{h^2}, \quad b = \frac{-2}{h^2}, \quad c = \frac{1}{h^2},$$

so the approximation has the form:

$$f''(x) = \frac{f(x) - 2f(x + h) + f(x + 2h)}{h^2} + (b + 8c)\frac{h^3}{3!}f'''(x) + \text{H.O.T.}$$

and the leading term in the error can now be calculated to be $hf'''(x)$.

7.3 Derivative Formulae Using Interpolants

Another way to obtain derivative formulae makes use of interpolation. An interpolating polynomial of the highest order possible is first constructed using the points in the stencil, then the derivative of this polynomial, which can be readily computed, is used as an approximation for the derivative of the original function. For an example, let us look at a stencil of three points, $\{x_0 = x - h, x_1 = x, x_2 = x + h\}$; this will allow us to obtain a quadratic polynomial $p_2(x)$ that can be easily written in Lagrange form:

$$p_2(x) = f_0 \frac{(x - x_1)(x - x_2)}{(x_0 - x_1)(x_0 - x_2)} + f_1 \frac{(x - x_0)(x - x_2)}{(x_1 - x_0)(x_1 - x_2)} + f_2 \frac{(x - x_0)(x - x_1)}{(x_2 - x_0)(x_2 - x_1)},$$

where, for convenience, we have used the notation $f_0 = f(x_0)$ etc., introduced in the discussion of interpolation. The first derivative of this polynomial with respect to x is:

$$\begin{aligned}
p_2'(x) &= f_0 \frac{[(x - x_2) + (x - x_1)]}{(x_0 - x_1)(x_0 - x_2)} + f_1 \frac{[(x - x_2) + (x - x_0)]}{(x_1 - x_0)(x_1 - x_2)} \\
&\quad + f_2 \frac{[(x - x_0) + (x - x_1)]}{(x_2 - x_0)(x_2 - x_1)}.
\end{aligned} \tag{7.4}$$

To obtain a centered formula for the derivative of the interpolated function $f(x)$ at the base point $x = x_1$, one can now set $x = x_1$ to obtain:

$$\begin{aligned}
p_2'(x_1) &= f_0 \frac{(x_1 - x_2)}{(x_0 - x_1)(x_0 - x_2)} + f_1 \frac{[(x_1 - x_2) + (x_1 - x_0)]}{(x_1 - x_0)(x_1 - x_2)} \\
&\quad + f_2 \frac{(x_1 - x_0)}{(x_2 - x_0)(x_2 - x_1)},
\end{aligned}$$

whereupon, taking into consideration that $x_1 - x_0 = h$ etc. the formula becomes:

$$f'(t_1) \approx p_2'(x_1) = f_0 \frac{-h}{2h^2} + f_1 \frac{[(-h) + h]}{(h)(-h)} + f_2 \frac{h}{(2h)(h)} = \frac{f_2 - f_0}{2h},$$

which can be seen to be the same centered formula obtained initially for the first derivative.

This approach may seem more complicated, but it does have the advantage that several approximations for a particular derivative can be obtained from the same polynomial. For example, one can set $x = x_0$ in equation (7.4) to obtain the forward formula, or set $x = x_2$ to obtain the backward-formula. Furthermore, approximations to the second derivative can also be obtained by taking one more derivative on the polynomial interpolant to yield:

$$p_2''(x) = f_0 \frac{2}{(x_0 - x_1)(x_0 - x_2)} + f_1 \frac{2}{(x_1 - x_0)(x_1 - x_2)} + f_2 \frac{2}{(x_2 - x_0)(x_2 - x_1)},$$
$$(7.5)$$

a relationship which again produces the same central derivative formula for $x = x_1$, namely:

$$f''(x_1) \approx p_2''(x_1) = f_0 \frac{2}{2h^2} + f_1 \frac{2}{(-h^2)} + f_2 \frac{2}{2h^2}.$$

Nevertheless, accounting for the error of the formulae is more complicated in this framework. For example, notice that the second derivative formula for the quadratic polynomial in equation (7.5) naturally does not contain the variable x any longer. Thus, one can legitimately use the same formula for the second derivative at all the three points in the stencil. However, while for the stencil midpoint the formula is second-order accurate:

$$f''(x_1) = \frac{f_0 - 2f_1 + f_2}{h^2} + \mathcal{O}(h^2),$$

the same approximation is only first-order accurate for the stencil endpoints, i.e.

$$f''(x_0) = \frac{f_0 - 2f_1 + f_2}{h^2} + \mathcal{O}(h).$$

7.4   Errors in Numerical Differentiation

Although conceptually straightforward, numerical differentiation is not without problems. These arise from the finite precision available for the representation of a floating-point number on a digital computer. Thus, while the analysis suggests that the error should decrease with the step size, it may happen that the linear combination of function values in the numerator of some particular formula is below machine precision. In that case the numerator is evaluated and stored with an error, which can be much exacerbated by the division with the denominator, especially if that involves a higher power of Δx.

Figure 7.1 shows the error in the numerical evaluation of the first derivative of $\sin(x)$ at $x = 0$ using the $\mathcal{O}(\Delta x^2)$ central approximation. For this function, the error does behave as expected for a large range of step sizes; in the logarithmic plot, the curve obtained is very close to a straight line with slope two, while further decreasing Δx makes the error vanish. A different behavior is however experienced for example for the function $f(x) = e^x/(1 + 2x^2)$. For this latter function, the error decreases with Δx only within some range, after which the trend switches gear and an unexpected increase in error becomes manifest, as seen in figure 7.2. This problem is exacerbated in the central approximation of the second derivative, as can be seen in figure 7.3. Numerical differentiation is an ill-conditioned operation and should be used carefully.

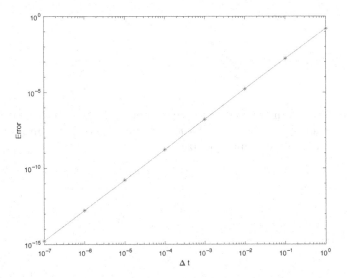

Figure 7.1: Error in the first derivative computed with the central formula for $f(x) = \sin(x)$. Error decreases as predicted by analysis.

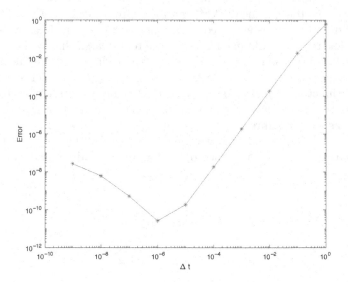

Figure 7.2: Error in the first derivative computed with the central formula for $f(x) = e^x/(1 + 2 * x^2)$. Rounding error becomes dominant below a certain step size.

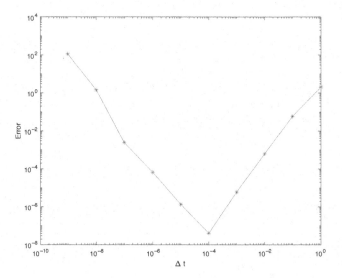

Figure 7.3: Error in the second derivative computed with the central formula for $f(x) = e^x/(1 + 2 * x^2)$.

7.5 Richardson Extrapolation

Formula extrapolation is a very useful technique that can be used to reduce the truncation error in both numerical differentiation and numerical integration when the form of the truncation error is known. In the numerical differentiation context it is usually known as *Richardson extrapolation*. While the approach will be presented in a general context, the discussion will focus afterward around the development of this method for the central first derivative formula (7.2). Upon taking into consideration more terms in the Taylor series expansion, the truncation error in this formula can be written, for a fixed value of x, as:

$$\frac{df}{dx} = \frac{f(x+h) - f(x-h)}{2h} + k_2 h^2 + k_4 h^4 + \ldots, \tag{7.6}$$

where the coefficients k_2, k_4 etc., depend only on the function $f(x)$ and on x. All the odd-order powers of h in the error disappear in this central approximation due to cancellation effects from points on either side of the base point. The fraction on the right-hand side in the above expression is the numerical approximation to the derivative, which will be denoted by $N(h)$ since it depends on the chosen step size h. For the most general case, let us suppose that one wants to approximate a quantity Q (in the case at hand $Q = df/dx$) using a formula $N(h)$ that depends on the step size. The truncation error will usually contain terms in all powers of h, so the approximation can be written:

$$Q = N(h) + k_1 h + k_2 h^2 + k_3 h^3 + k_4 h^4 + \ldots, \tag{7.7}$$

a form which highlights the fact that a possible improvement in the approximation could be achieved by eliminating the first-order term $k_1 h$, then the second-order term $k_2 h^2$ and so on. In order to achieve this goal, consider the approximation obtained by halving the step size:

$$\begin{aligned} Q &= N\left(\frac{h}{2}\right) + k_1\left(\frac{h}{2}\right) + k_2\left(\frac{h}{2}\right)^2 + k_3\left(\frac{h}{2}\right)^3 + \ldots \\ &= N\left(\frac{h}{2}\right) + k_1\left(\frac{h}{2}\right) + k_2\left(\frac{h^2}{4}\right) + k_3\left(\frac{h^3}{8}\right) + \ldots \end{aligned} \tag{7.8}$$

In this equation the quantity $N(h/2)$ represents the new numerical approximation with the smaller step size, which can justifiably be expected to be more accurate than $N(h)$. In the context of extrapolation, the two approximations $N(h)$ and $N(h/2)$ are used in conjunction to obtain an even more accurate result. To see how this occurs, consider multiplying (7.8) by a factor of two and subtracting (7.7) to obtain:

$$2Q - Q = 2\left(N\left(\tfrac{h}{2}\right) + k_1\left(\tfrac{h}{2}\right) + k_2\left(\tfrac{h^2}{4}\right) + k_3\left(\tfrac{h^3}{8}\right) + \ldots\right) -$$
$$\left(N(h) + k_1 h + k_2 h^2 + k_3 h^3 + \ldots\right) = \tag{7.9}$$
$$\left(2N\left(\tfrac{h}{2}\right) - N(h)\right) + k_2\left(\tfrac{h^2}{2} - h^2\right) + k_3\left(\tfrac{h^3}{4} - h^3\right) + \ldots$$

It is useful to now add an extra element to our notation in order to highlight the order of a numerical approximation. Thus, let

$$N_1(h) = N(h),$$
$$N_2(h) = 2N\left(\frac{h}{2}\right) - N(h),$$

be the two numerical approximations that can be obtained by using the procedure above. It follows from equation (7.9) that the two original approximations, when combined, produce a more accurate, $\mathcal{O}(h^2)$, formula for Q:

$$Q = N_2(h) - \frac{k_2}{2}h^2 - \frac{3k_3}{4}h^3 \ldots \tag{7.10}$$

The procedure is now followed recursively. Using the last equation with a step size $h/2$ leads to:

$$Q = N_2\left(\frac{h}{2}\right) - \frac{k_2}{2}\left(\frac{h}{2}\right)^2 - \frac{3k_3}{4}\left(\frac{h}{2}\right)^3 \ldots$$
$$= N_2\left(\frac{h}{2}\right) - \frac{k_2}{8}h^2 - \frac{3k_3}{32}h^3 - \ldots \tag{7.11}$$

One can now eliminate the next larger, $\mathcal{O}(h^2)$, error term by combining (7.10) with the latter equation to obtain:

$$4Q - Q = 3Q = 4N_2\left(\frac{h}{2}\right) - N_2(h) + \frac{3k_3}{4}\left(-\frac{h^3}{2} + h^3\right) + \ldots \tag{7.12}$$

whence:

$$Q = \frac{4}{3}N_2\left(\frac{h}{2}\right) - \frac{1}{3}N_2(h) + \mathcal{O}(h^3) = N_3(h) + \mathcal{O}(h^3), \tag{7.13}$$

and the process can be continued by subsequent divisions of the step size.

Let us now turn to the special case of the second-order central difference formula in equation (7.6). In this case, the truncation error contains only even powers of h, so the error elimination procedure will take the form in equation (7.13); that is, a multiplication by a factor of four will be needed upon halving the step size. A straightforward way to construct these successive approximations is then to start by computing the approximations from the original formula for a sequence of mesh sizes obtained by halving the intervals as:

$$E_{n,1} = N\left(\frac{h}{2^{n-1}}\right), \quad n = 1, 2, \ldots$$

The truncation error can then be expected to be of the order $\mathcal{O}(h^{2m})$ for the approximation $E_{n,m}$, which can be built recursively:

$$E_{n,m+1} = \frac{4^m}{4^m - 1}E_{n,m} - \frac{1}{4^m - 1}E_{n-1,m} \tag{7.14}$$

A MATLAB function that implements these calculations is shown in script (7.1). The following table shows the results obtained by using it on the function $f(x) = e^x/(1 + 2x^2)$ at the point $x = 0$ starting with $h = 0.25$.

TABLE 7.4: Richardson extrapolation for $f(x) = e^x/(1 + 2x^2)$.

h	$E_{i,1}$	$E_{i,2}$	$E_{i,3}$
0.25	0.898177	-	-
0.125	0.972224	0.996906	-
0.0625	0.992894	0.999784	0.999976

MATLAB Script 7.1: Richardson Extrapolation

```
function [val] = richardson(f, x, h, M)
  % evaluates the first derivative of f at x
  % Richardson extrapolation starting with h as coarsest levell
  val(1,1) = ( feval(f, x + h) - feval(f, x - h) ) / 2 / h;
  for i = 1:(M-1)
    h = h / 2;
    val(i+1,1) = ( feval(f, x + h) - feval(f, x - h) ) / 2 / h;
    for k = 1:i
      denom = 4 ^ k - 1;
      v = val(i+1,k);
      val(i+1,k+1) = v + ( v - val(i,k) ) / denom;
    end
  end
end
```

7.6 MATLAB® Built-in Functions

The basic tool that MATLAB offers for approximation of derivatives is `diff`. This function takes a vector of values x of length N and returns a vector of length $N - 1$ that contains the difference between successive values in x. Thus, in MATLAB notation, if dx = diff(x), then dx(1) = x(2) - x(1). The sample code below uses this built-in function to compute forward approximations to the first derivative of the function e^x on a grid with variable spacing.

MATLAB Script 7.2: Forward first derivative with diff

```
x = [0, 0.1, 0.24, 0.37];
y = exp(x);
dx = diff(x);
dy = diff(y);
dydx = dy ./ dx
```

The MATLAB function `gradient` extends the functionality of `diff` to functions of several variables. An example of its use for both the univariate and bivariate case is shown below.

MATLAB Script 7.3: Approximation using MATLAB's gradient

```
%1D gradient, fixed step size
h = 0.25;
x = 0:h:4.0;
y = exp(x);
dydx = gradient(y, h);
%2D gradient, fixed step size
h = 0.1;
x = -2:h:2;
y = x';
f = x .* y .^ 2;
[fx, fy ] = gradient(f);
```

7.7 Exercises

Ex. 1 Consider the function

$$f(x) = \cosh(x) = \frac{e^x + e^{-x}}{2}.$$

Evaluate the first and second derivatives at the point $x = 0$ using the second-order finite difference formulae in table 7.2. Use two step sizes, $h_1 = 0.5$ and $h_2 = 0.1$. Do the calculations using a hand calculator, rounding off every intermediate result to four decimal places.

Ex. 2 Consider the function

$$f(x) = \frac{e^x}{1 + x + 0.5x^2}.$$

Evaluate the error in its first derivative $f'(0)$, computed using the central first derivative formula, for a step size ranging from $\Delta x = 0.1$ to $\Delta x = 10^{-7}$. Decrease the value of the step Δx by a factor of ten for each successive evaluation. Plot the error versus the step Δx using a logarithmic scale on both axes. Repeat the same calculations for the second derivative, also computed with the centered formula. You may find that for some values of Δx the error is zero; when using a logarithmic scale, this may lead to trouble when plotting. To avoid this problem while still being able to see how the error evolves, you can just set the error to the machine accuracy **eps** whenever it is lower than **eps**.

Ex. 3 Write a MATLAB code that calculates the first and second derivatives of a given function on a grid $a = x_0 < x_1 < x_2 \ldots < x_n = b$ with constant grid spacing $x_{j+1} - x_j = h$ for $j = 0, 1, \ldots, (n-1)$ to second order of accuracy in h for all grid points. Plot the calculated values against the exact derivatives of the function $f(x) = x \sin(x)$ for $a = 0$, $b = 2\pi$ and $n = 50$.

Ex. 4 Use Richardson extrapolation to approximate the first derivative of the function

$$f(x) = \frac{e^x}{1 + 2x^2}$$

at the points $x = 0$ and $x = 1$. Use three mesh sizes defined by $h = h_1 = 0.4$, $h_2 = 0.2 = h_1/2$ and $h_3 = 0.1 = h_2/2$. Start by calculating by hand the approximate values $E_{1,1}$, $E_{2,1}$ and $E_{3,1}$, then extrapolate to find $E_{3,3}$. Perform all calculations by rounding off to four decimal places. Then write a MATLAB code that implements these calculations for the more general case of an arbitrary number of divisions of the step size. That amounts to allowing the second index m for the approximation $E_{n,m}$ to take values in some range $1, 2, \ldots, M$, where M is provided upon input. The range for n necessarily follows. Organize your code as a MATLAB script that calls a MATLAB function. The latter should accept M as input and compute the entries $E_{n,m}$ that can be stored in an output matrix.

Ex. 5 Using the Taylor series method outlined in section 7.2, find a formula that approximates the values for $f'(x)$ and $f''(x)$, with the highest possible accuracy, using a linear combination of these four values: $f(x)$, $f(x + h)$, $f(x + 2h)$, $f(x + 3h)$.

Ex. 6 Using the Taylor series method outlined in section 7.2, set up a 4×4 linear system for the coefficients c_i that would produce a third-order accurate finite difference approximation to $f'(x)$ based on 4 equally spaced points, of the form:

$$f'(x_0) \approx D_3 f(x_0)$$
$$= c_1 f(x_0 + h) + c_0 f(x_0) + c_{-1} f(x_0 - h) + c_{-2} f(x_0 - 2h) + \mathcal{O}(h^3).$$

Verify that the solution of your linear system gives rise the approximation rule

$$D_3 f(x_0) = \frac{1}{6h} \left[2f(x_0 + h) + 3f(x_0) - 6f(x_0 - h) + f(x_0 - 2h) \right].$$

Ex. 7 Given $h > 0$, find the coefficients A, B, and C in the following finite-difference approximation such that the result is exact if $f(x)$ is a polynomial of degree three or less, regardless of h:

$$D^2 f(x) \equiv \frac{Af(x + h) + Bf(x) + Cf(x - h)}{h^2} \approx f''(x).$$

Next, consider the function $f(x) = e^x$. Using the fact that the above finite-difference approximation gives exact values for polynomials of degree no more than three, obtain the leading-order error term of this differentiation rule for $f(x)$.

Ex. 8 Use the interpolation polynomial, as shown in section 7.3, to find a formula that approximates the value of the first derivative of the function $f(x)$ at all the stencil points $x_0 = x - h$, $x_1 = x$, $x_2 = x + h$ and $x_3 = x + 2h$.

Ex. 9 As discussed in Chapter 6, the interpolation polynomial can be written in different forms, with some of these being more suitable for certain tasks. Explore the use of the Newton's divided difference form of the interpolation polynomial for approximation of the derivative. Use the same stencil as in the above problem and obtain the approximations thus defined for both the first and the second derivatives at the first two points in the stencil.

Ex. 10 Extrapolation can be used to increase the accuracy of estimates for a variety of approximations that depend on a step size. Find a suitable one-step extrapolation formula that can be used to eliminate the highest-order term in the forward approximation to the first derivative presented in the first line in table 7.2.

Chapter 8

Numerical Optimization

Optimization is a rather general term which, in a technical sense, is closely related to finding minima or maxima of functions of one or more variables. The process has become known as *optimization* after numerical methods started being used extensively in technological design. In this context, optimization usually denotes minimization of a *cost function*, or rather functional, which encodes the effect that design parameters have on the performance of the final product. Let us remark from the very beginning that it is sufficient to have a method that finds the minima, since to obtain the maxima of a function one can search instead for the minima of the negative of the function to be maximized.

8.1   The Need for Optimization Methods

In principle, to find the extrema of a function, one can recast the problem as finding the zeros of the derivatives. Thus, one can instead solve an equation or a system thereof, as appropriate. Unfortunately, in many areas, and in particular in more involved technological processes, it is usually difficult to obtain explicit expressions for the derivatives of the function to be minimized. Take for example the process of designing a passenger aircraft. To make it simple, consider that the function, to be maximized in this case, is the lift over drag ratio. The lift is the vertical force that keeps the aircraft in the air, the drag the force opposing motion. Both are connected to the shape of the aircraft, in particular the shape of the wings, which in a simplified setting is governed by a set of *design parameters*, which we can think of as representing the spatial coordinates of a set of points on the wing. The design process roughly involves in this scenario a couple of steps: calculation of the airflow around the aircraft for the particular design parameters under consideration (requires the solution of a system of partial differential equations) and calculation of the ensuing lift and drag and their ratio from this resulting airflow. This ratio is the cost function to be maximized and requires an integral of the pressure obtained in the previous step over the surface of the plane. Clearly computing the derivative of the cost function with respect to the design parameters is not an easy endeavor, as the function itself is the output of a large numerical simulation. While techniques like *automatic differentiation* have been developed to this purpose, the case where no derivatives are immediately available is the one that concerns us here.

Optimization is known as *unconstrained* when the extrema of the cost function are sought without any restriction on the design parameters. This is rarely the case in engineering design for example, where concerns about structural stability and weight, as well as financial imperatives, must be taken into account. To use the same aircraft example, drag can be minimized if the airplane is reduced to a flat, sheet-of-paper-like shape; but this would be useless for the purpose of transporting passengers, therefore constraints are imposed on the design parameters to enforce a functional shape of the wings and fuselage. This, in turn, results in a more difficult optimization process known as *constrained optimization*.

8.2   Line Search Methods

Our discussion starts with the unconstrained minimization problem. The methods discussed in this section work under the assumption that only the value of the function to be minimized, a function of only one variable denoted by $f(x)$, is readily available. That is, the first derivatives do not need to be known or evaluated for the methods in this section to work. Nevertheless, an extra assumption on the function $f(x)$ is needed. First, a function $f : \mathbb{R} \to \mathbb{R}$ has a *minimum* at a point $x_m \in [a, b]$ if $f(x) \geq f(x_m)$ for all $x \in [a, b]$. Furthermore, the function is *unimodal* on $[a, b]$ if $f(x)$ is strictly decreasing for $x \leq x_m$ and strictly increasing for $x \geq x_m$; in other words, there is a single local minimum in the interval $[a, b]$. This property, depicted in figure 8.1,

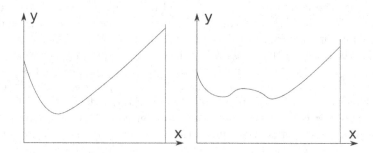

Figure 8.1: A function that is unimodal (left) contrasted with a multimodal function (right).

allows us to construct a sequence of intervals of decreasing size that contain the minimum.

8.2.1 Bracketing

The method of *bracketing* a minimum is very similar to the bisection method for finding the root of a nonlinear equation of the form $f(x) = 0$ for a continuous function $f(x)$ within a suitably chosen bracket $[a, b]$. The difference resides in the fact that, for bisection, one can obtain a suitable bracket for the root by just comparing the sign of the two values of the function $f(x)$ at the endpoints of the interval. By contrast, one can only be sure that the bracket $[a_1, b_1]$ contains a minimum only after comparing function values at a minimum of three points. Consider an additional point m_1 inside the bracket, i.e. such that $a_1 < m_1 < b_1$; if $f(m_1) < f(a_1)$ and $f(m_1) < f(b_1)$, then the interval $[a_1, b_1]$ contains a minimum of $f(x)$. Sometimes this is expressed by stating that the triplet of points (a_1, m_1, b_1) brackets the minimum, a useful but not very accurate expression. Once the initial bracket is known, the method uses the unimodal property of the function and proceeds as follows:

▶ Choose an extra point $c \in [a_1, m_1]$;

▶ Construct the next bracket; if $f(c) < f(m_1)$, then the triplet (a_1, c, m_1) brackets the minimum so the next bracket is $[a_2 = a_1, b_2 = m_1]$ and $m_2 = c$; if, on the other hand, $f(c) > f(m_1)$, then the next bracket is $[a_2 = c, b_2 = b]$ and $m_2 = m_1$.

These two situations are illustrated on the right and left plots, respectively, in figure 8.2.

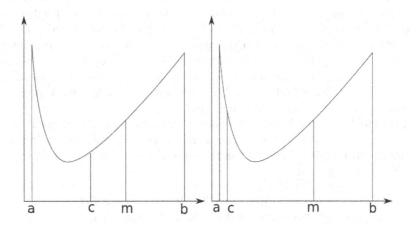

Figure 8.2: Choosing an optimization bracket for $f(x)$.

8.2.2 Golden section search

The *golden section search* method offers a way to make the bracketing method described above very efficient. It reduces the size of the bracketing interval by a constant factor upon each subdivision, while at the same time keeping the number of function evaluations low through the reuse of previously computed values. The method starts with the choice of an initial triplet (a_1, m_1, b_1) that brackets the minimum, and offers a way to choose the new point c. Since $m_1 \in (a_1, b_1)$, start by defining the two ratios $r, s \in (0, 1)$ by:

$$r = \frac{m_1 - a_1}{b_1 - a_1}; \quad s = 1 - r = \frac{b_1 - m_1}{b_1 - a_1}.$$

One can say that the position of m_1 is a fraction r of the distance from a_1 to b_1 away from a_1, and search for the next point c an additional fraction t beyond m_1, that is:

$$t = \frac{c - m_1}{b_1 - a_1}.$$

The next bracketing segment will therefore either be of length $r + t$ or of length $s = 1 - r$ relative to the current one. The golden section search asks that the relative size of the new interval be the same in either case, so that $t = s - r = 1 - 2r$. Moving to the new subinterval and additionally requiring the same ratio to be obtained leads to:

$$r = \frac{t}{s},$$

which, combined with the previous relationship for r, produces the quadratic equation $r^2 - 3r + 1 = 0$, with the only appropriate solution $r \approx 0.382$. Thus, if the points m_1 and c are chosen at relative positions r and $s = 1 - r$ within the original interval (a_1, b_1), the next bracketing triplet (a, m_1, c) and (m_1, c, b) will both be $1 - r$ times the original length $b_1 - a_1$. After N bracketing choices are made, the original interval is reduced $s^N \approx 0.618^N$ times. The quantity $\phi = 1 + s$ is sometimes known as the *golden ratio*, whence the name of the method. The golden search method converges linearly to the local minimum if the function to be optimized is unimodal.

Let us look at a simple example of how to organize the calculations involved in performing one step of the golden search method to find the minimum of the unimodal function $f(x) = x^2$ on the interval $(a_1 = -2, b_1 = 5)$. First, one obtains the two points:

$$m_{11} = a_1 + r(b_1 - a_1) = 0.674; \quad f(m_{11}) = 0.4542,$$
$$m_{12} = a_1 + s(b_1 - a_1) = 2.326; \quad f(m_{12}) = 5.4103.$$

Since $f(m_{11}) < f(m_{12})$, the next interval for the search is $(a_2 = a_1, b_2 = m_{12})$, in this case $(-2, 2.326)$. If the opposite were the case, that is for situations when $f(m_{11}) > f(m_{12})$, the subinterval to be retained would be (m_{11}, b_1).

The MATLAB script (8.1) shows a straightforward implementation of the golden search method for the function $f(x) = \sin x$, which is unimodal on the interval $[\pi, 5\pi/2]$. The code in the script is not efficient; it doesn't make use of the fact that only one point needs to be recomputed upon subsequent subdivisions of the bracketing interval. The efficient implementation is left as an exercise.

MATLAB Script 8.1: Golden Search Method

```
clear all;
f = @(x) sin(x) ;
xL = pi;
xR = 5*pi/2;
r = 0.382;
T = 10^(-5); % tolerance
for n = 1:200
  xM1 = xL + (xR - xL)*r;
  xM2 = xL + (xR - xL)*(1-r);
  if f(xM1) < f(xM2)
    xR = xM2;
  else
    xL = xM1;
  end
  % stop once the bracket is small enough
  if (xR - xL ) < T
    break;
  end
end
% show number of iterations performed
n % output: 28
% since the interval is small now, one can use the midpoint
% as an approximation for the position of the minimum
disp( (xR + xL) / 2)  % output: 4.7124
% this is the exact minimum
disp(3*pi/2)  % output: 4.7124
```

Because the bracketing interval is reduced by a factor s at each iteration, it follows that the midpoint of the last subinterval is at most a distance $s^n(b-a)/2$ from the true minimum after n subdivisions. Therefore, just like for the bisection method, using this relationship one can easily find the number of iterations required for a specified tolerance.

8.3 Successive Parabolic Interpolation

A way to make additional use of the information provided by the three function values that are computed in the simple bracketing of a minimum is by fitting a parabola to these data points. The minimum of the parabola will, in turn, provide an approximation for the actual minimum of the function under investigation. The ensuing method, known as *successive parabolic interpolation*, usually converges much faster than the line search methods described above. This does come with a penalty, however, in that the three points that are chosen to define the parabola may turn out to be on the same line. In this case the method fails to produce an approximation to the minimum at that particular stage and needs to revert to the safer line search methods.

Suppose again that the minimum is in the interval $[a, b]$, with a third point m inside the interval. Then the parabola that interpolates the function $f(x)$ using these three points is given by its Lagrange form:

$$p_2(x) = f(a)\frac{(x-m)(x-b)}{(a-m)(a-b)} + f(m)\frac{(x-a)(x-b)}{(m-a)(m-b)} + f(b)\frac{(x-a)(x-m)}{(b-a)(b-m)}$$

The derivative of this function is therefore:

$$\frac{dp_2(x)}{dx} = f(a)\frac{(x-m)+(x-b)}{(a-m)(a-b)} + f(m)\frac{(x-a)+(x-b)}{(m-a)(m-b)}$$

$$+f(b)\frac{(x-a)+(x-m)}{(b-a)(b-m)}$$

so that the point x_m for which the derivative is zero is given by:

$$x_m = m + \frac{1}{2}\frac{[f(a)-f(m)](b-m)^2 - [f(b)-f(m)](m-a)^2}{[f(a)-f(m)](b-m) + [f(b)-f(m)](m-a)}$$

Unless the initial point m is itself the minimum of $p_2(x)$, the new x_m satisfies $f(x_m) < f(m)$. Therefore, the new bracketing triplet can be chosen to be (a, x_m, m) if $x_m < m$ or (x_m, m, b) in the opposite case. The iterative process can be stopped when the size of the bracketing interval is under a certain tolerance.

8.4 Optimization Using Derivatives

8.4.1 Univariate functions

For functions of only one variable for which the derivatives are known, finding the minimum of the function f is equivalent to solving for the root of the derivative f'. Therefore, in this case Newton's method can be used with an initial guess x_0 for the minimum, successive updates being given by

$$x_{k+1} = x_k - \frac{f'(x_k)}{f''(x_k)}.$$

As previously discussed, Newton's method converges quadratically if the initial guess is adequate, but is not completely safe. A combination of methods that takes advantage of the fast convergence of Newton's method while reverting to safer bracketing (like the golden section search) when it fails is a common solution for univariate unconstrained optimization.

8.4.2 Steepest descent for multivariate functions

For functions of several variables, an important information is offered by the gradient vector: it points in the direction of maximum local increase in the function. This information may, in turn, be used to obtain successive approximations for extremum points. Let $f : \mathbb{R}^n \to \mathbb{R}$ be a function of n variables $x_1, x_2, \ldots, x_n$; then the *gradient vector* at a point $\mathbf{x}^0$ is given by:

$$\nabla f|_{\mathbf{x}^0} = \begin{bmatrix} \frac{\partial f}{\partial x_1} \\ \frac{\partial f}{\partial x_2} \\ \vdots \\ \frac{\partial f}{\partial x_n} \end{bmatrix}_{\mathbf{x}^0} \tag{8.1}$$

With a proper choice for the initial guess $\mathbf{x}^0$, the *steepest descent* method produces the next approximation for the minimum using the following iterative definition:

$$\mathbf{x}^{k+1} = \mathbf{x}^k - s_k \nabla f|_{\mathbf{x}^k}, \ k = 0, 1, \ldots \tag{8.2}$$

where the parameter $s_k > 0$ is a step size that determines how far one goes in the direction opposite to the local gradient (hence toward the minimum). The value of s_k can in turn be estimated by solving an additional one-dimensional optimization problem for the function

$$g(s_k) = f(\mathbf{x}^k - s_k \nabla f|_{\mathbf{x}^k}), \tag{8.3}$$

where $\mathbf{x}^k$ is of course kept fixed. The steepest descent method allows one to always make progress toward the minimum, as long as the gradient is not the zero vector. However, the successive points $\mathbf{x}^k$ may zigzag back and forth and convergence can be very slow, the convergence rate being at most linear in the general case.

8.4.3 Solving linear systems

For the case of a symmetric and positive definite matrix of order n, the problem of finding the solution of a linear system can be restated as finding the minimum of a function of n variables. To show how this happens, the case $n = 2$ will be discussed here for simplicity. The system to be solved in this case is:

$$\begin{aligned} a_{11}x + a_{12}y &= b_1 \\ a_{21}x + a_{22}y &= b_2 \end{aligned}$$

or equivalently $A\mathbf{x} = \mathbf{b}$, where we suppose that the system matrix

$$A = \begin{pmatrix} a_{11} & a_{12} \\ a_{21} & a_{22} \end{pmatrix}$$

is symmetric, that is $a_{12} = a_{21} = a$ in this case. Consider now the function $F(x, y) = F(\mathbf{x})$ defined by:

$$F(x,y) = \frac{1}{2}\mathbf{x}^T A\mathbf{x} - \mathbf{x}^T\mathbf{b} = \frac{1}{2}\left[a_{11}x^2 + 2axy + a_{22}y^2\right] - b_1 x - b_2 y \qquad (8.4)$$

The function $F(x, y)$ has a critical point when $A\mathbf{x} = \mathbf{b}$, as can be easily confirmed by calculating its first derivatives and requiring that they equal zero simultaneously:

$$\frac{\partial F}{\partial x} = a_{11}x + ay - b_1$$
$$\frac{\partial F}{\partial y} = ax + a_{22}y - b_2$$

The fact that the critical point of $F(x, y)$ is a minimum is guaranteed by the positive definiteness of the system matrix A.

Let's consider for example the solution of the two-by-two linear system

$$A\mathbf{x} = \mathbf{b}, \quad \text{where } A = \begin{pmatrix} 2 & 1 \\ 1 & 3 \end{pmatrix}, \ \mathbf{b} = \begin{bmatrix} 3 \\ -1 \end{bmatrix}$$

The gradient of the function $F(x, y) = x^2 + xy + 1.5y^2 - 3x + y$ is then given by $\nabla F(x, y) = [2x + y - 3, x + 3y + 1]^T$. Such a function of two variables can also be easily visualized using the plotting capabilities in MATLAB. The first figure produced by script 8.2 clearly shows the surface is shaped like a bowl and has a minimum.

MATLAB Script 8.2: Surface and Contour Visualization

```
close all;
clear all;
F = @(x,y) x.^2 + x.*y + 1.5*y.^2 - 3*x + y;
x = linspace(-6,6);
y = linspace(-6,6);
[X,Y] = meshgrid(x,y);
FXY = F(X,Y);
figure(1)
surf(x,y,FXY)
xlabel('X')
ylabel('Y')
zlabel('F')
figure(2)
contour(X,Y,FXY,'k'), hold on
grid on
quiver(0,0,3,-1,'k')
plot(2,-1, '*k')
```

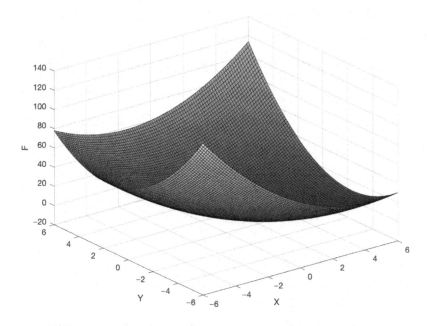

Figure 8.3: Surface representation of the function F.

Starting from the initial guess $\mathbf{x}^0 = \mathbf{0}$, the local maximum increase direction is therefore $\nabla F(0,0) = [-3, 1]^T$; notice that this is also the residual vector $\mathbf{r}^0 = A\mathbf{x}^0 - \mathbf{b}$ at the point $\mathbf{x}^0$. Therefore the next iterate will be $\mathbf{x}^1 = \mathbf{x}^0 - s_0[-3, 1]^T$. The remaining task is to determine the step size s_0. To this end, notice that $\mathbf{x}^1 = [3s_0, -s_0]^T$, so replacing this in the definition of the function leads to $F(\mathbf{x}^1) = \frac{15}{2}s_0^2 - 10s_0$. Therefore, the extremum of this function, which only depends on s_0 now, is seen to be obtained when $dF/ds_0 = 15s_0 - 10 = 0$, hence $s_0 = \frac{10}{15}$. This in turn leads to the new iterate $\mathbf{x}^1 = [0, 0]^T - (10/15)[-3, 1]^T$. The new iterate is therefore $[2, -0.666]^T$, which is closer to the exact solution of this system, $\mathbf{x}^* = [2, -1]^T$. While in a real-life problem the exact solution will not be known, to quantify convergence one can again compare the norm of the residual vectors for successive approximations. In this case the norm of the initial residual vector is $||\mathbf{r}^0|| = 3.1623$, while for the next approximation $||\mathbf{r}^1|| = 1.0541$. Figure 8.4, the second figure created by script 8.2, shows the contours of the function together with the negative gradient direction at the initial guess point and the actual position of the root.

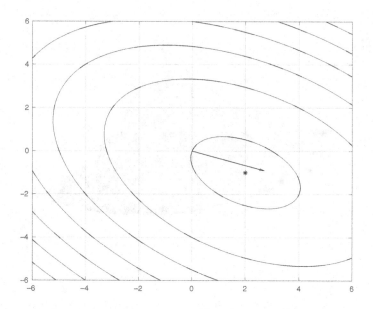

Figure 8.4: Contours of the function $F(x, y)$ and the steepest descent direction at point $(0, 0)$.

We now extend the quadratic function $F(\mathbf{x}) = \frac{1}{2}\mathbf{x}^T A\mathbf{x} - \mathbf{x}^T\mathbf{b}$ in (8.4) to a more general setting, for which $\mathbf{x} = (x_1, x_2, \ldots, x_n)^T \in \mathbb{R}^n$, $\mathbf{b} = (b_1, b_2, \ldots, b_n)^T \in \mathbb{R}^n$, A is an $n \times n$ symmetric positive definite matrix. The function $F(\mathbf{x}) \in \mathbb{R}$ remains a scalar function. Denoting the entries of matrix A by $\{a_{ij}\}$, $i, j = 1, 2, \ldots, n$, recall that A is SPD if:

1. $a_{ij} = a_{ji}$,

2. $\mathbf{x}^T A\mathbf{x} > 0$.

Under this setting, the optimization problem

$$\min_{\mathbf{x}\in\mathbb{R}^n} F(\mathbf{x}) = \min_{\mathbf{x}\in\mathbb{R}^n}\left(\frac{1}{2}\mathbf{x}^T A\mathbf{x} - \mathbf{x}^T\mathbf{b}\right) \tag{8.5}$$

has a unique minimum at a point $\mathbf{x}^*$, where $\nabla F(\mathbf{x}^*) = 0$. Since

$$\nabla F(\mathbf{x}) = A\mathbf{x} - \mathbf{b},$$

the minimizer $\mathbf{x}^*$ solves the linear system

$$A\mathbf{x} - \mathbf{b} = \mathbf{0}. \tag{8.6}$$

This means that solving the optimization problem (8.5) is equivalent to solving the linear system (8.6), and the steepest descent method (8.2), written as

$$\mathbf{x}^{k+1} = \mathbf{x}^k - s_k\nabla F(\mathbf{x}^k) = \mathbf{x}^k + s_k\mathbf{r}_k, \quad k = 0, 1, \ldots, \tag{8.7}$$

where $\mathbf{r}_k = -(A\mathbf{x}^k - \mathbf{b})$, is indeed an iterative method for solving systems of linear equations, $A\mathbf{x} = \mathbf{b}$. As we mentioned earlier in (8.3), the scalar s_k is chosen by solving the adjacent minimization problem

$$\min_{s \in \mathbb{R}} F(\mathbf{x}^k - s\nabla F(\mathbf{x}^k)) = \min_{s \in \mathbb{R}} F(\mathbf{x}^k + s\mathbf{r}_k). \qquad (8.8)$$

Note that

$$F(\mathbf{x}^k + s\mathbf{r}_k) = \frac{1}{2}(\mathbf{x}^k + s\mathbf{r}_k)^T A(\mathbf{x}^k + s\mathbf{r}_k) - (\mathbf{x}^k + s\mathbf{r}_k)^T \mathbf{b}$$

$$= \left(\frac{1}{2}(\mathbf{x}^k)^T A\mathbf{x}^k - (\mathbf{x}^k)^T \mathbf{b} \right) + s(\mathbf{r}_k^T A\mathbf{x}^k - \mathbf{r}_k^T \mathbf{b}) + \qquad (8.9)$$

$$\frac{1}{2}s^2 \mathbf{r}_k^T A\mathbf{r}_k.$$

Thus, the minimum of (8.8) occurs when

$$\frac{dF(\mathbf{x}^k + s\mathbf{r}_k)}{ds} = \mathbf{r}_k^T A\mathbf{x}^k - \mathbf{r}_k^T \mathbf{b} + s(\mathbf{r}_k^T A\mathbf{r}_k) = 0.$$

Solving the above equation for s gives the step size at iteration k,

$$s_k = \frac{\mathbf{r}_k^T \mathbf{r}_k}{\mathbf{r}_k^T A\mathbf{r}_k}. \qquad (8.10)$$

The steepest descent method for solving the linear system $A\mathbf{x} = \mathbf{b}$ is summarized in algorithm 1.

Algorithm 1 Steepest descent method for solving $A\mathbf{x} = \mathbf{b}$.

1: Initial guess $\mathbf{x}^0$
2: Compute $\mathbf{r}_0 = \mathbf{b} - A\mathbf{x}^0$
3: **for** $k = 1, 2, \ldots$ **do**
4:     $\mathbf{z}_{k-1} = A\mathbf{x}^{k-1}$
5:     $s_{k-1} = \dfrac{\mathbf{r}_{k-1}^T \mathbf{r}_{k-1}}{\mathbf{r}_{k-1} \mathbf{z}_{k-1}}$
6:     $\mathbf{x}^k = \mathbf{x}^{k-1} + s_{k-1}\mathbf{r}_{k-1}$
7:     $\mathbf{r}_k = \mathbf{r}_{k-1} - s_{k-1}\mathbf{z}_{k-1}$
8:     **if** $\|\mathbf{r}_k\| < \epsilon$ **then**
9:         break
10:    **end if**
11: **end for**

Notice that algorithm 1 is made efficient in Step 7 by computing

$$\mathbf{r}_k = \mathbf{b} - A\mathbf{x}^k$$

$$= \mathbf{b} - A(\mathbf{x}^{k-1} + s_{k-1}\mathbf{r}_{k-1})$$

$$= \mathbf{r}_{k-1} - s_{k-1}\mathbf{z}_{k-1},$$

because by using this alternative formula, the algorithm reuses the vector $\mathbf{z}_{k-1}$ and avoids performing an additional matrix-vector multiplication.

The Conjugate Gradients Method

In practice, the steepest descent method is oftentimes used as an algorithm for nonlinear optimization problems. It is not used for solving linear systems because of its possibly slow convergence. Instead, the conjugate gradient method, a method closely related to the steepest descent method, is one of the most popular iterative methods for solving linear systems, $A\mathbf{x} = \mathbf{b}$ when the coefficient matrix A is SPD. The conjugate gradient method was introduced by Hestenes and Stiefel in 1952, but it took decades for the scientific computing community to fully appreciate its potential.

Unlike the steepest descent method that uses a fixed direction vector $\mathbf{r}_k = \mathbf{b} - A\mathbf{x}^k$ in each iteration, the conjugate gradient method uses a set of nonzero direction vectors $\{\mathbf{v}_1, \mathbf{v}_2, \ldots, \mathbf{v}_n\}$ that satisfy the *A-orthogonal condition*

$$\mathbf{v}_i^T A\mathbf{v}_j = 0, \quad \text{if } i \neq j. \tag{8.11}$$

Using the A-orthogonal direction vectors produces the step size

$$s_k = \frac{\mathbf{v}_k^T \mathbf{r}_{k-1}}{\mathbf{v}_k^T A\mathbf{v}_k} \tag{8.12}$$

and the update rule for the solution is

$$\mathbf{x}^k = \mathbf{x}^{k-1} + s_k \mathbf{v}_k, \quad k = 1, 2, \ldots. \tag{8.13}$$

To construct the A-orthogonal direction vectors, one uses $\mathbf{r}_k$ to generate $\mathbf{v}_{k+1}$ by setting

$$\mathbf{v}_{k+1} = \mathbf{r}_k + t_k \mathbf{v}_k. \tag{8.14}$$

Here, t_k must be chosen so that the A-orthogonal condition (8.11) is satisfied. Multiplying A to (8.14) from the left, one has

$$A\mathbf{v}_{k+1} = A\mathbf{r}_k + t_{k-1} A\mathbf{v}_k.$$

Since the A-orthogonal condition (8.11) requires

$$\mathbf{v}_k^T A\mathbf{v}_{k+1} = \mathbf{v}_k^T A\mathbf{r}_k + t_k \mathbf{v}_k^T A\mathbf{v}_k = 0,$$

one obtains

$$t_k = -\frac{\mathbf{v}_k^T A\mathbf{r}_k}{\mathbf{v}_k^T A\mathbf{v}_k} = -\frac{\mathbf{r}_k^T A\mathbf{v}_k}{\mathbf{v}_k^T A\mathbf{v}_k}, \tag{8.15}$$

since A is symmetric. Now from equation (8.13), it follows that

$$\mathbf{b} - A\mathbf{x}^k = \mathbf{b} - A\mathbf{x}^{k-1} - s_k A\mathbf{v}_k \quad \text{or} \quad \mathbf{r}_k = \mathbf{r}_{k-1} - s_k A\mathbf{v}_k$$

Thus

$$\mathbf{r}_k^T \mathbf{r}_k = \mathbf{r}_{k-1}^T \mathbf{r}_k - s_k \mathbf{r}_k^T A \mathbf{v}_k = -s_k \mathbf{r}_k^T A \mathbf{v}_k, \tag{8.16}$$

since $\mathbf{r}_i^T \mathbf{r}_j = 0$, for $i \neq j$; thus $\mathbf{r}_{k-1}^T \mathbf{r}_k = 0$. This implies

$$\mathbf{r}_k^T A \mathbf{v}_k = -\frac{1}{s_k} \mathbf{r}_k^T \mathbf{r}_k. \tag{8.17}$$

Using mathematical induction, one can show that

$$\mathbf{r}_k^T \mathbf{v}_j = 0, \quad \text{for each } j = 1, 2, \ldots, k. \tag{8.18}$$

We will leave the proof for the above equation as an exercise. Using equations (8.14) and (8.18), one can rewrite (8.12) as follows:

$$s_k = \frac{\mathbf{v}_k^T \mathbf{r}_{k-1}}{\mathbf{v}_k^T A \mathbf{v}_k} = \frac{(\mathbf{r}_{k-1} + t_{k-1} \mathbf{v}_{k-1})^T \mathbf{r}_{k-1}}{\mathbf{v}_k^T A \mathbf{v}_k} = \frac{\mathbf{r}_{k-1}^T \mathbf{r}_{k-1}}{\mathbf{v}_k^T A \mathbf{v}_k}, \tag{8.19}$$

since $\mathbf{v}_{k-1}^T \mathbf{r}_{k-1} = 0$ by (8.18). This implies

$$\mathbf{v}_k^T A \mathbf{v}_k = \frac{1}{s_k} \mathbf{r}_{k-1}^T \mathbf{r}_{k-1} \tag{8.20}$$

Substituting equations (8.17) and (8.20) into (8.15), one obtains

$$t_k = \frac{\mathbf{r}_k^T \mathbf{r}_k}{\mathbf{r}_{k-1}^T \mathbf{r}_{k-1}}. \tag{8.21}$$

Collecting equations (8.13), (8.14), (8.19) and (8.21) together, the conjugate gradient method is summarized in algorithm 2.

Let us now see how the conjugate gradient method works by considering the following linear system:

$$
\begin{aligned}
3x_1 - x_2 + x_3 &= 0, \\
-x_1 + 6x_2 + 2x_3 &= 9, \\
x_1 + 2x_2 + 7x_3 &= -2.
\end{aligned}
$$

Let $\mathbf{x} = (x_1, x_2, x_3)^T$. Written in matrix-vector form, the system is $A\mathbf{x} = \mathbf{b}$, where the coefficient matrix A and the right-hand-side vector $\mathbf{b}$ are:

$$
A = \begin{bmatrix} 3 & -1 & 1 \\ -1 & 6 & 2 \\ 1 & 2 & 7 \end{bmatrix} \quad \text{and} \quad \mathbf{b} = \begin{bmatrix} 0 \\ 9 \\ -2 \end{bmatrix}.
$$

The system matrix A is an SPD matrix (the proof is left as an exercise). The solution of the linear system is $\mathbf{x}^* = (1, 2, -1)^T$. Let the initial guess be $\mathbf{x}^0 = (0, 0, 0)^T$. $\mathbf{r}_0 = \mathbf{b} - A\mathbf{x}^0 = \mathbf{v}_1 = (0, 9, -2)^T$. The following computations show, up to the four decimals, each term in algorithm 2 for the first three iterations ($k \leq 3$):

Algorithm 2 Conjugate gradient method for solving $Ax = b$.

1: Initial guess $\mathbf{x}^0$
2: Compute $\mathbf{r}_0 = \mathbf{b} - A\mathbf{x}^0$
3: $\mathbf{v}_1 = \mathbf{r}_0$
4: **for** $k = 1, 2, \ldots$ **do**
5:     $\mathbf{z}_k = A\mathbf{v}_k$
6:     $s_k = \dfrac{\mathbf{r}_{k-1}^T \mathbf{r}_{k-1}}{\mathbf{v}_k^T \mathbf{z}_k}$
7:     $\mathbf{x}^k = \mathbf{x}^{k-1} + s_k \mathbf{v}_k$
8:     $\mathbf{r}_k = \mathbf{r}_{k-1} - s_k \mathbf{z}_k$
9:     **if** $\|\mathbf{r}_k\| < \epsilon$ **then**
10:         break
11:     **end if**
12:     $t_k = \dfrac{\mathbf{r}_k^T \mathbf{r}_k}{\mathbf{r}_{k-1}^T \mathbf{r}_{k-1}}$
13:     $\mathbf{v}_{k+1} = \mathbf{r}_k + t_k \mathbf{v}_k$
14: **end for**

$k = 1$

$\mathbf{z}_1 = (-11, 50, 4)^T$; $s_1 = 0.1923$; $\mathbf{x}^1 = (0, 1.7308, -0.3846)^T$; $\mathbf{r}_1 = (2.1154, -0.6154, -2.7692)^T$; $t_1 = 0.1473$; $\mathbf{v}_2 = (2.1154, 0.7105, -3.0639)^T$.

$k = 2$

$\mathbf{z}_2 = (2.5718, -3.9802, -17.9107)^T$; $s_2 = 0.2178$; $\mathbf{x}^2 = (0.4608, 1.8855, -1.0520)^T$; $\mathbf{r}_2 = (1.5552, 0.2516, 1.1321)^T$; $t_2 = 0.3006$; $\mathbf{v}_3 = (2.1910, 0.4651, 0.2112)^T$.

$k = 3$

$\mathbf{z}_3 = (6.3191, 1.0222, 4.5999)^T$; $s_3 = 0.2461$; $\mathbf{x}^3 = (1.000, 2.000, -1.000)^T$.

The conjugate gradient method finds the exact solution, up to the machine precision, after only three iterations ($k = 3$). The number of iterations is equal to the dimension of the linear system ($n = 3$). In general, the conjugate gradient method is guaranteed to converge in at most n iterations for a linear system with an SPD coefficient matrix of order n.

The MATLAB implementation of algorithm 2 for this example is listed in MATLAB script 8.3. The MATLAB built-in conjugate gradient method is included in the script for comparison.

MATLAB Script 8.3: Conjugate gradient algorithm

```
clear all;
A= [3 -1 1; -1 6 2; 1 2 7];
b=[0;9;-2];
maxit=100; %maximum number of iteration
epsilon=1e-12; %stopping criterion
x=[0;0;0]; %initial guess
r_1=b-A*x;
v=r_1;

for k=1:maxit
    z=A*v;
    s=(r_1'*r_1)/(v'*z);
    x=x+s*v;
    r=r_1-s*z;
    if (norm(r,2)< epsilon)
        break
    end
    t= (r'*r)/(r_1'*r_1);
    v=r+t*v;
    r_1=r;
end
k
exact_x=[1;2;-1]
x

% Matlab built-in preconditioned conjugate gradient method. The
    preconditioners M1 & M2 are not assigned.

tol=1e-12;
M1=[];
M2=[];
x0=[0;0;0];
[Matlab_x,flag,relres,iter,resvec] = pcg(A, b,
    tol,maxit,M1,M2,x0)
```

8.5 Linear Programming

Our study of constrained optimization starts with linear programming. A linear program is simply a constrained optimization problem that features a linear objective function and linear constraints. The constraints may include both equalities and inequalities. The *simplex method*, introduced by George Dantzig in the 1940s, is an iterative procedure based on Gauss-Jordan elimination that tackles this problem. This method is in wide use for solving problems arising from economics studies. The following example illustrates the use of the simplex method for solving a standard maximization problem.

Profit function analysis

A company manufactures two products, X and Y, on machines A and B. The company will profit \$3/unit for product X and \$2/unit for product Y. Manufacturing one unit of product X requires two minutes on machine A and 2 minutes on machine B. Manufacturing one unit of product Y requires three minutes on machine A and one minute on machine B. There are twelve minutes of time available on machine A and eight minutes of time available on machine B for every half-hour. The interest is in knowing how many units of each product the company should produce every thirty minutes in order to maximize its profit.

To set up a linear program, assume that the company produces x units of product X and y units of product Y for every 30 minutes. The objective function P is the profit in this timeframe, thus the problem can be stated as:

$$\text{Maximize } P = 3x + 2y \tag{8.22}$$

subject to the constraints

$$2x + 3y \le 12,$$
$$2x + y \le 8, \tag{8.23}$$
$$x \ge 0, y \ge 0.$$

The first step of the simplex method is to replace the system of inequality constraints (8.8) with a system of equality constraints. This can be achieved by using nonnegative variables called *slack variables*. For example, one may add a nonnegative α to the left-hand side of the inequality

$$2x + 3y \le 12$$

to compensate for the difference and thus obtain the equality

$$2x + 3y + \alpha = 12.$$

Similarly, convert $2x + y \le 8$ into $2x + y + \beta = 8$. The system of linear constraints is then

$$2x + 3y + \alpha = 12,$$
$$2x + y + \beta = 8.$$

To obtain a linear system of equations for the optimization problem, one rewrites the objective function (8.8) in the form $-3x - 2y + P = 0$. Finally, the standard maximization problem becomes:

$$\begin{aligned} 2x + 3y + \alpha \quad\quad &= 12, \\ 2x + y \quad + \beta \quad &= 8, \\ -3x - 2y \quad\quad +P &= 0 \end{aligned} \tag{8.24}$$

where x, y, α, β, P are all nonnegative variables. Notice that the ensuing system (8.24) has five unknowns but only three equations. One may therefore solve for three of the unknowns in terms of the other two. This results in infinitely many solutions to the system in terms of two parameters. The *feasible solutions* to the problem are the solutions, for which x, y, α and β are all nonnegative.

To proceed, consider the following augmented matrix associated with system (8.24) is

$$
\begin{array}{ccccc}
x & y & \alpha & \beta & P \\
\left[\begin{array}{ccccc|c}
2 & 3 & 1 & 0 & 0 & 12 \\
2 & 1 & 0 & 1 & 0 & 8 \\
-3 & -2 & 0 & 0 & 1 & 0
\end{array}\right]
\end{array}
\tag{8.25}
$$

Notice that each of the columns associated with the variables α, β and P is a unit column. The variables associated with unit columns are known as **basic variables**; all other variables are **nonbasic variables**. The simplex method proceeds by choosing pivot elements and applying row operations to reach the optimal solution. The main steps of the method are summarized in algorithm 3.

Algorithm 3 The simplex method

1: Set up the initial augmented matrix associated with the linear program.
2: Examine all entries in the last row to the left of the vertical line.
3: **if** all entries are nonnegative, the optimal solution has been reached. **then**
4:     Proceed to step 9.
5: **else if** there are one or more negative entries, the optimal solution has not been reached. **then**
6:     Proceed to step 8.
7: **end if**
8: Convert columns associated with nonbasic variables into unit columns.
  (a) Select the pivot element: the pivot element is the intersection of the pivot column and the pivot row. The pivot column is the column that contains the most negative entry in the last row to the left of the vertical line. The pivot row is obtained by first dividing every positive entry in the pivot column into its corresponding entry in the column to the right of the vertical line. The pivot row is then chosen as the row corresponding to the smallest ratio.

  (b) Apply row operations based on the pivot element to convert the pivot column into a unit column; row operations include: (i) multiply all entries in a row by a nonzero constant and (ii) replace one row by a sum of itself and a multiple of another row.
9: Determine the optimal solutions: the value of the variable heading each unit column is given by the entry in the column to the right of the vertical line in the row containing the 1.

For the maximization problem introduced above, (8.8) and (8.8), the first step of the simplex method yields the augmented matrix (8.25). Since there are negative entries in the last row, we select the first column as our pivot column (most negative entry). Also the second row is our pivot row because $8/2 = 4 < 6 = 12/2$. Thus the pivot element is 2.

$$
\rightarrow
\begin{array}{ccccc}
x & y & \alpha & \beta & P \\
\end{array}
\left[\begin{array}{ccccc|c}
2 & 3 & 1 & 0 & 0 & 12 \\
\boxed{2} & 1 & 0 & 1 & 0 & 8 \\
-3 & -2 & 0 & 0 & 1 & 0
\end{array}\right]
\tag{8.26}
$$

Now one must apply row operations to convert the first column into a unit vector with the unit entry 1 located at the position of the pivot element.

$$
\begin{array}{c}
R_1 \leftarrow R_2 \times (-2) + R_1 \\
R_2 \leftarrow 1/2 \times R_2 \\
R_3 \leftarrow 3 \times R_2 + R_3
\end{array}
\begin{array}{ccccc}
x & y & \alpha & \beta & P \\
\end{array}
\left[\begin{array}{ccccc|c}
0 & \boxed{2} & 1 & -1 & 0 & 4 \\
1 & 1/2 & 0 & 1/2 & 0 & 4 \\
0 & -1/2 & 0 & 3/2 & 1 & 12
\end{array}\right]
\tag{8.27}
$$

For matrix (8.27), the second column is the pivot column, while the first row is the pivot row. Thus the pivot element is 2 in the second column. Continue by performing the necessary row operations to obtain the final augmented matrix:

$$
\begin{array}{c}
R_1 \leftarrow 1/2 \times R_1 \\
R_2 \leftarrow R_1 \times (-1/2) + R_2 \\
R_3 \leftarrow R_1 \times 1/2 + R_3
\end{array}
\begin{array}{ccccc}
x & y & \alpha & \beta & P \\
\end{array}
\left[\begin{array}{ccccc|c}
0 & 1 & 1/2 & -1/2 & 0 & 2 \\
1 & 0 & -1/4 & 3/4 & 0 & 3 \\
0 & 0 & 1/4 & 5/4 & 1 & 13
\end{array}\right]
\tag{8.28}
$$

Since there are no more negative entries in the last row, we have reached the optimal solution. The optimal solution is $x = 3$, $y = 2$ and $P = 13$.

Let us now turn our attention to the mechanisms behind the simplex method. Recall matrix (8.25); x and y are nonbasic variables and α, β and P are basic variables. One can represent the basic variables by the nonbasic variables in the form:

$$
\begin{aligned}
\alpha &= 12 - 2x - 3y, \\
\beta &= 8 - 2x - y, \\
P &= 3x + 2y.
\end{aligned}
\tag{8.29}
$$

Letting the nonbasic variables be zero, $x = 0$ and $y = 0$ in (8.29), one clearly obtains $P = 0$. This solution is called a *basic solution*. To increase the value of P from the basic solution, one can either increase the value of x or the value of y. Notice that it is more profitable to increase x. This selection corresponds to the choice of pivot column, i.e. the most negative entry in the last row of

(8.25). The next question is how much can one increase x while holding $y = 0$. To answer the question, let $y = 0$ in (8.29) and get

$$\alpha = 12 - 2x,$$
$$\beta = 8 - 2x. \tag{8.30}$$

Since $\alpha, \beta \geq 0$, the increase of x cannot exceed $\frac{8}{2} = 4$, as seen from the second equation of (8.30). This condition corresponds to the choice of the pivot row, i.e. the row that contains the smallest ratio of a positive entry in the pivot column dividing its corresponding entry in the column to the right side of the vertical line.

If one thus chooses $x = 4$ while fixing $y = 0$, one obtains the solution

$$x = 4, \quad y = 0, \quad \alpha = 4, \quad \beta = 0, \quad P = 12. \tag{8.31}$$

By definition, (8.31) is a basic solution with nonbasic variables $y = 0$ and $\beta = 0$. This basic solution suggests that the basic variables are x, α and P, which means that the columns of the augmented matrix under x, α and P are unit columns. The augmented matrix after one iteration of the simplex method, shown in (8.27), satisfies the requirement. In fact, if one writes down the system of equations for basic variables using nonbasic variables in (8.27), one has:

$$x = 4 - \frac{1}{2}y - \frac{1}{2}\beta,$$
$$\alpha = 4 - 2\,y + \beta, \tag{8.32}$$
$$P = 12 + \frac{1}{2}y - \frac{3}{2}\beta.$$

It is clear that the basic solution (8.31) can be obtained by letting the nonbasic variables $y = 0$ and $\beta = 0$ in (8.32).

The simplex method for minimization problems

Every maximization problem is associated with a minimization problem, and vice versa. The associated problem is known as the **dual problem** of the given problem. To apply the simplex method for a minimization problem, we find the dual maximization problem associated with the given problem, and then solve the dual problem with algorithm 3. The process is presented in the next example.

Consider the minimization problem:

Minimize the objective function $\quad C = 3x + 2y,$

subject to the constraints

$$2x + y \geq 6,$$
$$x + y \geq 4.$$

An augmented matrix corresponding to the given problem is

$$\begin{array}{cc} x & y \\ \begin{bmatrix} 2 & 1 & | & 6 \\ 1 & 1 & | & 4 \\ \hline 3 & 2 & | & 0 \end{bmatrix} \end{array}$$

The augmented matrix corresponding to the dual maximization problem is simply the transpose of the one above,

$$\begin{array}{cc} u & v \\ \begin{bmatrix} 2 & 1 & | & 3 \\ 1 & 1 & | & 2 \\ \hline 6 & 4 & | & 0 \end{bmatrix} \end{array}$$

The dual maximization problem can thus be stated:

Maximize the objective function $\qquad P = 6u + 4v,$

subject to $\qquad\qquad\qquad 2u + v \le 3,$

$$u + v \le 2,$$

$$u \ge 0,\ v \ge 0.$$

The connection between the solution of the given problem and that of the dual problem was established by John von Neumann. The fundamental theorem of duality given by von Neumann guarantees that the objective functions of both problems attain the same optimal value.

The MATLAB function **linprog** provides an option to solve a minimization linear programing problem by using the simplex method, with constraints of the form $Ax \le b$. For the maximization problem (8.8) and (8.8), the dual minimizing has the objective function $C = -3x - 2y$ and the same set of constraints. A MATLAB implementation by using the provided simplex method is:

MATLAB Script 8.4: The simplex method

```
clear all;
close all;
f=[-3, -2];
A=[2,3 ; 2, 1];
b=[12, 8];
Aeq=[];
beq=[];
lb=[];
ub=[];
format short
options = optimoptions('linprog','Algorithm','dual-simplex')
[x,fval,exitflag,output] = linprog(f,A,b,Aeq,beq,lb,ub,options)
```

Running this MATLAB script produces the output:

```
Optimal solution found.
x =
    3.0000
    2.0000
fval =
    -13
```

The minimum is given by fval $= -13$, which means that the maximum is $P = 13$. This solution is thus consistent with the optimal solution found previously.

8.6   Constrained Nonlinear Optimization

If the objective functions or the constraints are nonlinear functions, one may be able to solve the constrained optimization problems using Lagrange multipliers[18]. For an example, let us consider the following constrained optimization problem with two variables:

$$\text{Minmize} \quad f(x, y) = 2x^2 + y^2 + 2, \tag{8.33}$$

$$\text{subject to} \quad x^2 + 4y^2 - 4 = 0. \tag{8.34}$$

The objective function is here $2x^2 + y^2 + 2 = f(x, y)$; it is also useful to introduce the notation $g(x, y) = x^2 + 4y^2 - 4$ for the constraint function. f is differentiable and has a local extremum value on the curve $g(x, y) = 0$ at a point $P = (a, b)$. One can parameterize this curve near P using the vector form; to this end, denote by $r(t)$ the position along the curve, where t is the curve parameter and $r(t)$ is differentiable. One can conveniently choose $r(t_0) = P$. As we vary t along g, the rate of change of f is given by the chain rule:

$$\frac{d}{dt} f(r(t)) \Big|_{t=t_0} = \nabla f_P \cdot r'(t) \Big|_{t=t_0} = 0.$$

This shows that ∇f_P is orthogonal to the tangent vector $r'(t_0)$ to the curve $g(x, y) = 0$. The gradient ∇g_P is also orthogonal to the tangent vector $r'(t_0)$ to the curve $g(x, y) = 0$ at P. One can thus conclude that ∇f_P is parallel to ∇g_P. Hence there exists a scalar λ so that

$$\nabla f(a, b) = \lambda \nabla g(a, b). \tag{8.35}$$

The scalar λ is called a *Lagrange multiplier*.

Noting that $\nabla f(x, y) = (4x, 2y)$ and $\nabla g(x, y) = (2x, 8y)$, it follows from (8.35) and the constraint that the coordinates of the point P must obey the system of equations

$$4a = \lambda(2a), \tag{8.36}$$

$$2b = \lambda(8b), \tag{8.37}$$

$$a^2 + 4b^2 - 4 = 0. \tag{8.38}$$

The solutions of (8.36) are $a = 0$ or $\lambda = 2$. If $a = 0$, (8.38) implies that $b = \pm 1$. The solutions of (8.37) are $b = 0$ or $\lambda = \frac{1}{4}$. If $b = 0$, (8.38) implies that $a = \pm 2$. Therefore, we conclude that the solutions to the system of equations, (8.36) − (8.38), are either

$$a = 0, b = \pm 1, \lambda = \frac{1}{4}, \tag{8.39}$$

or

$$a = \pm 2, b = 0, \lambda = 2. \tag{8.40}$$

The extreme values of f occur at $(a, b) = (0, \pm 1)$ and $(a, b) = (\pm 2, 0)$. Since $f(0, \pm 1) = 3$ and $f(\pm 2, 0) = 10$, we see that the minimum value of f subject to the constraint is 3, which occurs at $(0, 1)$ and $(0, -1)$. Notice that the final result does not use the value of λ.

Notice that one can define a Lagrange function using (8.35),

$$\mathcal{L}(x, y) = f(x, y) - \lambda g(x, y)$$

for which it follows that

$$\nabla \mathcal{L}(x, y)\Big|_{(x,y)=(a,b)} = 0.$$

Sometimes people prefer to define this function in the following way instead:

$$\mathcal{L}(x, y) = f(x, y) + \tilde{\lambda} g(x, y),$$

which is obviously an equivalent expression if $\tilde{\lambda} = -\lambda$.

The MATLAB function `fmincon` solves constrained nonlinear optimization problems by using a so-called interior-point algorithm and returns estimated Lagrange multipliers in a structure. The MATLAB implementation below solves the nonlinear minimization problem (8.33) and (8.34) by using `fmincon`.

MATLAB Script 8.5: Constrained nonlinear optimization

```
clear all;
close all;
lb=[];
ub=[];
A =[];
b= [];
Aeq=[];
beq=[];
fun = @(x) 2*x(1)^2 +x(2)^2+2;
x0=[0,0];
format short
[x,fval,exitflag,output,lambda] =
    fmincon(fun,x0,A,b,Aeq,beq,lb,ub,@mycon);
x
fval
lambda

function [c,ceq] = mycon(x)
c=[];
ceq = x(1)^2+4*x(2)^2-4;
end
```

The script specifies an initial search point, $x_0 = [0,0]$. The solution to the optimization problem is then found to be:

```
x =
    0.0000    1.0000
fval =
    3.0000
```

Since MATLAB uses the alternative formulation, the returned estimated Lagrange multiplier is -0.25, instead of $+0.25$ in our analysis. The local optimal solution found by MATLAB is consistent with (8.39).

8.7   MATLAB® Built-in Functions

Apart from the functions already mentioned above, the MATLAB function fminbnd can be used to find the minimum of a function of one variable on a bounded interval. To use it, define the function to be minimized and present it as an input argument to fminbnd, together with the bracket wherein the search is to be conducted. The example below exemplifies the procedure.

MATLAB Script 8.6: Using fminbnd with default options

```
close all;
clear all;
%function to be minimized
f = @(x) exp(x) ./ x.^2;

%search bracket
xL = 0.1;
xR = 5.0;

% find the minimum
xMin = fminbnd(f, xL, xR) %output: xMin =  2.0000
```

As in many other MATLAB built-in functions, the progress of the iterative process can be displayed by sending additional optional inputs. In addition, the minimum value of the function can also be obtained as an additional output.

MATLAB Script 8.7: Using fminbnd with extra display options

```
close all;
clear all;
%function to be minimized
f = @(x) exp(x) ./ x.^2;

%search bracket
xL = 0.1;
xR = 5.0;

% display iteration table
options = optimset('Display','iter');
%compute minimum location and function value
[xMin, fVal] = fminbnd(f, xL, xR, options)
```

The function `fminsearch` performs a derivative-free search for the minimum of a function of several variables on an unbounded domain. The script below shows how to use the function to obtain the default output together with a plot of the evolution of the function value through the iterative process. A tabulated version of the progress of the iteration can be obtained with the same option as in the previous script. Furthermore, several options can be combined together within the same call to `optimset`.

MATLAB Script 8.8: Using fminsearch with plot options

```matlab
close all;
clear all;
%function to be minimized
f = @(x) x(1)^2 + x(1)*x(2) + 1.5 * x(2)^2 - 3 * x(1) + x(2);

%initial point
x0 = [0.0, 0.0];

% find the minimum
options = optimset('PlotFcns',@optimplotfval);
xMin = fminsearch(f, x0, options) %output: 1.99993  -0.99994
```

The function `fminunc` can be used in a similar way to `fminsearch` when used for derivative-free unconstrained optimization. It allows, however, the specification of the gradient of the function to be minimized. In most cases this will make the optimization problem converge much faster. Below is a script that uses this `fminunc`, the gradient of the function to be minimized being provided as an input. When the derivative or gradient is provided, other options to solve minimization problems obviously include `fzero` and `fsolve`.

MATLAB Script 8.9: Using fminunc with specified gradient

```matlab
%initial point
x0 = [0.0, 0.0];

% find the minimum
options =
    optimoptions('fminunc','SpecifyObjectiveGradient',true);
xMin = fminunc(@sysGrad, x0, options) %output: 2.0000  -1.0000

%function to be minimized
function [f,J] = sysGrad(x)
  f = x(1)^2 + x(1)*x(2) + 1.5 * x(2)^2 - 3 * x(1) + x(2);
  J = [ 2 * x(1) + x(2) - 3; x(1) + 3*x(2) + 1];
end
```

As seen above, constrained multivariable optimization can be performed in MATLAB with the `fmincon` function. In addition, the `lsqnonneg` function attempts to find the vector **x** that minimizes the norm of $\mathbf{r} = \mathbf{b} - A\mathbf{x}$, subject to the constraint that all the components of **x** be non-negative, $x_i \geq 0$.

MATLAB Script 8.10: Using lsqnonneg: to write

```
%initial point
x0 = [0.0, 0.0];

% find the minimum
options =
    optimoptions('fminunc','SpecifyObjectiveGradient',true);
xMin = fminunc(@sysGrad, x0, options) %output: 2.0000   -1.0000

%function to be minimized
function [f,J] = sysGrad(x)
 f = x(1)^2 + x(1)*x(2) + 1.5 * x(2)^2 - 3 * x(1) + x(2);
 J = [ 2 * x(1) + x(2) - 3; x(1) + 3*x(2) + 1];
end
```

In addition, because of the technological importance of optimization problems, MATLAB has a full toolbox dedicated to the task. Its exploration is left to the interested reader.

8.8 Exercises

Ex. 1 Is the function defined by $f(x) = x^2 - 4x + 3 - \ln(x)$ unimodal on the interval $x \in [1, 5]$? Plot the function to investigate its behavior, then perform two iterations of the golden search method. Do all calculations by hand, carrying four digits after the decimal point.

Ex. 2 In the discussion of bracketing in 8.2.1, the case $f(c) = f(m_1)$ was not covered. What is the appropriate subinterval choice in this case?

Ex. 3 Starting from the script 8.1, implement the golden search method in an efficient way to take advantage of the fact that only one of the two points inside a subinterval needs to be computed after the first iteration. Use your implementation to find the minimum of the same function $f(x) = x^2 - 4x + 3 - \ln(x)$ in the first problem above on the same interval $[1, 5]$ with a tolerance $T = 10^{-5}$. Print your first two iterations and verify your hand calculations for that problem.

Ex. 4 Write a MATLAB function that implements the successive parabolic interpolation method. Use your implementation to find the minimum of the function $f(x) = 0.6 - xe^{-x^2}$ on the interval $[0, 3]$, with a tolerance $T = 10^{-5}$.

Ex. 5 Find the point on the graph of the function $f(x) = x + 1$ that is closest to the point with coordinates $(x = 3, y = -3)$. Recast the problem as a minimization problem and use the golden section search method to search for the minimum with a tolerance $T = 10^{-6}$.

Ex. 6 Use both successive parabolic interpolation and the golden section search method to find the local minimum of the function $f(x) = 2xe^x$ on the interval $[-3, 0]$.

Ex. 7 Solve the linear system of equations:

$$\begin{cases} 3x + y - z & = & 4 \\ x + 5y + 2z & = & -1 \\ -x + 2y + 5z & = & 1 \end{cases}$$

using the steepest descent method starting from the usual initial guess $\mathbf{x}^0 = [0, 0, 0]^T$. For each iteration, print the iteration number k, the current approximation $\mathbf{x}^k$, and the norm of the residual at this current location. Start with $k = 0$ and produce a table with this information. Convergence should be declared when the norm of the residual is below 10^{-7}.

Ex. 8 Show that the matrix of the linear system in **Ex. 7** is SPD. Modify the MATLAB script 8.3 and solve the linear system with the conjugate gradients method. Similar to **Ex. 7**, let the initial guess be $\mathbf{x}^0 = [0, 0, 0]^T$ and convergence should be declared when the norm of the residual is below 10^{-7}.

Ex. 9 Compute the gradient $\nabla\phi(\mathbf{x})$ and the so-called Hessian matrix of second derivatives, $\nabla^2\phi(\mathbf{x})$, for the Rosenbrock function

$$\phi(\mathbf{x}) = 100(x_2 - x_1^2)^2 + (1 - x_1)^2.$$

Show that $\mathbf{x}^* = (1, 1)^T$ is the only local minimizer of this function, and that the Hessian matrix at that point is positive definite.

Ex. 10 Write a steepest descent algorithm in MATLAB for solving **Ex. 9**. Set the initial step length $\alpha_0 = 1$ and print the step length used at each iteration. First try the initial point $\mathbf{x}_0 = (1.2, 1.2)^T$ and then the more difficult starting point $\mathbf{x}_0 = (-1, 1.2)^T$.

Ex. 11 Repeat **Ex. 10** but use Newton's method.

Ex. 12 Solve the following maximization problem by using the simplex method described in Algorithm 3:

$$\text{Maximize } P = 2x + 2y + z$$

subject to the constraints

$$2x + y + 2z \leq 14,$$
$$2x + 4y + z \leq 26,$$
$$x + 2y + 3z \leq 28,$$
$$x \geq 0, \; y \geq 0, \; z \geq 0.$$

Ex. 13 Modify MATLAB script 8.4 to solve **Ex. 12**.

Ex. 14 Use Lagrange multipliers to find the maximum area of a rectangle whose four corner points, in counterclockwise order, are (x, y), $(-x, y)$, $(-x, -y)$ and $(x, -y)$ and is inscribed in the ellipse

$$\frac{x^2}{256} + \frac{y^2}{64} = 1.$$

Ex. 15 Modify MATLAB script 8.5 to solve **Ex. 14**.

Chapter 9

Numerical Quadrature

This chapter addresses the numerical approximation of integrals, a process that is also known as *numerical quadrature*. The term *quadrature*, to be understood here as *area evaluation*, apparently has its origin in ancient Greece, where the members of Pythagora's mathematical philosophy school thought that the only way to obtain the area of a geometrical figure was to produce a square with the same area. Needless to say, they were not very successful in their endeavor for the circular disk.

9.1 Basic Quadrature Formulae

Several of the most common methods may have been already encountered in Calculus classes, for example the trapezoidal method and Simpson's method. Both of these are based on first approximating the function to be integrated by a polynomial, using a predefined set of points, then integrating the resulting polynomial instead of the function. The methods that use this procedure are collectively known as Newton-Cotes methods, and can be further divided into two categories, closed- versus open-end methods.

Before describing these basic methods, it is useful to settle on some notational conventions. In this section we'll switch back to working with a continuous function $f(x)$, the definite integral of which between a and b will be denoted by

$$I[f; a, b] = \int_a^b f(x)dx.$$

If the function involved in the integral is clear from the context, it will not be indicated in the above notation, which will thus simplify to $I[a, b]$.

9.1.1 The trapezoidal method

The *trapezoidal method* or *trapezoidal rule* gets its name from the fact that the integral $I[f; a, b]$ is approximated by the area of a trapezoid obtained from the two vertical lines $x = a$ and $x = b$, the horizontal axis, and the straight line that joins the two points $(a, f(a))$ and $(b, f(b))$. This scenario is pictured

DOI: 10.1201/9780429262876-9

in figure 9.1. Therefore, the approximation takes the form:

$$I[f; a, b] \approx \frac{b - a}{2} (f(a) + f(b)) \tag{9.1}$$

The error involved in this approximation can be calculated starting from the result in theorem 6.2.2. In this particular case, if the function $f(x)$ together with its derivatives up to second order are continuous, it follows that:

$$f(x) = p_1(x) + \frac{f''(\xi)}{2!} R_2(x); \quad R_2(x) = (x - a)(x - b),$$

hence the desired integral is given by

$$I[f; a, b] = \int_a^b p_1(x)dx + \int_a^b \frac{f''(\xi)}{2!} R_2(x)dx,$$

where the first term on the right of the equal sign is the approximation and the second constitutes the error, which can be written

$$\int_a^b \frac{f''(\xi)}{2}(x - a)(x - b)dx = \frac{f''(\xi)(a - b)^3}{12},$$

whence, by denoting $b - a = h$, it follows that the error in the trapezoidal formula is given by:

$$I[f; a, b] = \frac{b - a}{2} (f(a) + f(b)) - \frac{f''(\xi)h^3}{12},$$

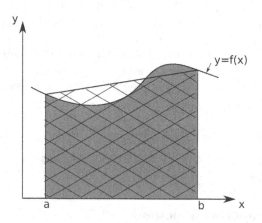

Figure 9.1: Trapezoidal rule approximation.

The approximation can obviously be improved by considering smaller sub-divisions. Let us therefore suppose that the interval $[a, b]$ is partitioned as

$a = x_0 < x_1 \ldots < x_n = b$, with a constant distance between two successive points, $x_{i+1} - x_i = \Delta x = h$ for $i = 0, 1, \ldots, n-1$. It will prove useful for the subsequent discussion to call a subdivision $[x_i, x_{i+1}]$ a *patch*; the value of the function at grid point x_i will also be denoted $f(x_i) = f_i$ for simplicity. Then the integral $I[a, b]$ can be expressed as:

$$I[a, b] = I[a, x_1] + I[x_1, x_2] + \ldots + I[x_{n-2}, x_{n-1}] + I[x_{n-1}, b],$$

where the trapezoidal approximation in equation (9.1) will be used on each patch, leading to the so-called *composite trapezoidal rule*:

$$I[a, b] \approx \frac{h}{2} \left(f_0 + 2f_1 + 2f_2 + \ldots + 2f_{n-1} + 2f_{n-1} + f_n \right).$$

To investigate the truncation error involved in this formula, one can start from the error involved in the simple rule on only one patch. Since the error on any patch will be $\mathcal{O}(h^3)$, the collective error on all the n patches will be $n\mathcal{O}(h^3)$. Taking into consideration the fact that, for a given interval $[a, b]$, the number of patches is related to the patch width by $n = (b-a)/h$, the collective error for the composite trapezoidal rule can be seen to be $(b-a)\mathcal{O}(h^2)$, hence $\mathcal{O}(h^2)$ because the interval length is a constant.

9.1.2   Simpson's method

Instead of a piecewise-linear approximation, the method usually referred to as *Simpson's method* uses a quadratic approximation built on two patches. For the simple Simpson's rule, a depiction of which is given in figure 9.2, a midpoint is inserted in between the two endpoints of the interval $[a, b]$. Using a notation similar to the above, let the distance between successive points be denoted by $\Delta x = h = (b-a)/2$. The function $f(x)$ is approximated by a quadratic $p_2(x)$, which can easily be written using the Lagrange form for the interpolant,

$$p_2(x) = f_0 \frac{(x - x_1)(x - x_2)}{(x_0 - x_1)(x_0 - x_2)} + f_1 \frac{(x - x_0)(x - x_2)}{(x_1 - x_0)(x_1 - x_2)} + f_2 \frac{(x - x_0)(x - x_1)}{(x_2 - x_0)(x_2 - x_1)},$$

This can in turn be integrated to obtain the following approximation to the integral:

$$I[a, b] = \int_a^b f(x)dx \approx \int_a^{b=a+2h} p_2(x)dx = \frac{h}{3} \left(f_0 + 4f_1 + f_2 \right). \tag{9.2}$$

Investigating the truncation error incurred by this approximation on the basis of theorem 6.2.2 alone, in this case written for a quadratic:

$$I[a, b] = \int_a^b p_2(x)dx + \int_a^{a+2h} \frac{f'''(\xi)R_3(x)}{6}dx$$

is somewhat misleading here, because the cubic remainder term integrates to zero. Therefore, one needs to consider higher-order terms in the local approximation of $f(x)$. It can easily be shown that the Taylor series for $f(x)$ has the form:

$$f(x) = p_2(x) + f'''(x_m)\frac{t(t^2 - h^2)}{6} + f''''(x_m)\frac{t^2(t^2 - h^2)}{24} + \ldots,$$

where $x_m = a + h$ is the midpoint of the interval $[a, b]$ and the notation $t = x - x_m$ was used for convenience. Notice that the third derivative term in the above expression is identical with the one in theorem 6.2.2, as should be expected. Upon integrating this expression, the third-derivative term on the right-hand side integrates to zero; this leaves the integral of the last term as the main contributor to the error, which therefore has the form:

$$I[a, b] = \frac{h}{3}(f_0 + 4f_1 + f_2) - \frac{f''''(\xi)h^5}{90} + \ldots.$$

This means that, although Simpson's method starts with a quadratic approximation, the result is also exact for a cubic. Such a gain of one order of accuracy in the approximation is due to the fact that the method uses an extra point in the middle of the interval, and is closely related to the extra order of accuracy achieved by centered differentiation formulae; the errors on the two subintervals to the left and right of the midpoint, respectively, have a different sign and thus cancel each other.

Similar to the composite trapezoidal rule, for a more accurate approximation one can again further subdivide the interval $[a, b]$. Suppose that this is done by partitioning it with the points $a = x_0 < x_1 \ldots < x_{n-1} < x_n = b$, with $x_{j+1} - x_j = h$ and n an even integer such that the total number of subintervals of length h is even. It is again useful to give these subintervals a name; while by no means a standard nomenclature, let us agree that such a couple of adjacent patches on which a single quadratic approximation is used will be called a *panel*. There will thus be K panels and $n = 2K$ patches involved in the approximation. The composite Simpson rule is then obtained by noticing that

$$\begin{aligned} I[a, b] \quad &= \quad I[a, x_2] + I[x_2, x_4] + \cdots + I[x_{n-4}, x_{n-2}] + I[x_{n-2}, b] \\ &\approx \quad \tfrac{h}{3}(f_0 + 4f_1 + 2f_2 + 4f_3 + 2f_4 + \ldots + 4f_{n-1} + f_n). \end{aligned}$$ (9.3)

An argument identical to the one used for the composite trapezoidal rule shows that, because the error for the simple Simpson's rule is $\mathcal{O}(h^5)$, the composite rule has a truncation error of the order $\mathcal{O}(h^4)$.

A sample code that uses Simpson's rule to compute the integral of the function $f(x) = xe^{-x^2}$ on the interval $[a = 0, b = 2]$ using twenty patches is given in MATLAB script 9.1. As usual for numerical computation, efficiency is again emphasized; notice how the code avoids repeated multiplication by four and two, respectively, in the two loops but performs only one multiplication instead when summing up the results. Also notice that, while the use of two separate loops was not mandatory, it does make the code more legible and easy to follow.

MATLAB Script 9.1: Simpson Method

```
clear all;       % clears all variables
f = @(x) x .* exp( - x .* x );
a = 0   ;     % left boundary
b = 2   ;     % right boundary
n = 20;
h = (b-a)/n;
x = a:h:b;

s1 = f(a) + f(b);
s2 = 0;  % for function values with weight 2
s4 = 0; % for function values with weight 4
for j = 2:2:length(x)-1
    s4 = s4 + f( x(j) );
end
for j = 3:2:length(x)-2
    s2 = s2 + f( x(j) );
end
approxInt = h/3*( s1 + 4*s4 + 2*s2 ) %prints computed value
```

There exists another method sometimes referred to as *Simpson's 3/8 rule*, which is obtained by approximating the function $f(x)$ with piecewise cubics defined on three patches. In order to use this method one therefore needs n to be a multiple of three. The composite formula obtained for Simpson's 3/8 rule has the same order of accuracy, $\mathcal{O}(h^4)$, as the usual Simpson's method, albeit a smaller constant multiplies the h^4 term; the method will not be discussed here any further.

9.1.3   Open-end versus closed-end methods

Both the trapezoidal rule and Simpson's rule ask for evaluation of the function values at both endpoints of the interval; such methods are known as closed-end quadrature methods. While most of the time this does not raise any issues, there are definite integrals for which this may not be feasible. As a straightforward example, consider the function $f(x) = 1/(2\sqrt{x})$ and the following integral:

$$\int_0^4 f(x)dx = \sqrt{x}\Big|_0^4 = 2,$$

which cannot be evaluated with either method because both require the value of the function at $x = 0$.

Open-end methods, on the other hand, avoid this problem by only requiring the value of the function to be integrated at points located inside the interval. The simplest such method is obtained by using again an even number of patches, just as for Simpson's rule; this time however, the function $f(x)$ is

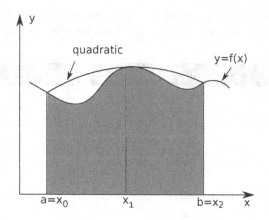

Figure 9.2: The integral of $f(x)$ and its Simpson's rule approximation by the area under a quadratic.

approximated on each panel by a constant, the value of which is set equal to the value of $f(x)$ at the midpoint of the panel. Thus, on the interval $[x_0, x_2]$, the function is approximated by $y(x) = f_1$, which leads to the following:

$$I[x_0, x_2] \approx 2hf_1$$

and the composite rule:

$$
\begin{aligned}
I[a, b] &= I[x_0, x_2] + I[x_2, x_4] + \ldots + I[x_{n-2}, x_n] \\
&\approx 2h\left(f_1 + f_3 + f_5 + \ldots f_{n-1}\right).
\end{aligned}
\tag{9.4}
$$

This approximation is known as the midpoint rule. Notice that, for a given subdivision, it involves half as many function evaluations as the composite trapezoidal rule. Nonetheless, it can be shown that this method has the same order of accuracy $\mathcal{O}(h^2)$ as the composite trapezoidal rule.

9.2   Gauss Quadrature

The Newton-Cotes formulae use a fixed set of points to interpolate the function that needs to be integrated. Gauss quadrature takes a different approach: the points used in approximating the function, as well as their relative weights in the approximation, are selected in such way as to maximize the accuracy of the approximation. To understand how this process works, one can start by contemplating the integral of a first-order polynomial of the form $f(x) =$

$Ax + B$:

$$\int_a^b (Ax + B)dx = A \int_a^b x\,dx + B \int_a^b 1\,dx.$$

Suppose now that one can compute the two quantities $\int_a^b x\,dx$ and $\int_a^b 1\,dx$ exactly. Since the integral of $f(x)$ is only a linear combination of these values, it follows that for such linear functions one can also compute $\int_a^b f(x)dx$ exactly by just using the proper linear combination of these simpler integrals.

This reasoning can be extended to polynomials of arbitrary degree. Any polynomial of degree n in x can be conveniently written as a linear combination of functions in the set of functions $\{1, x, x^2, \ldots, x^n\}$, collectively known as the *standard basis* for polynomials. Thus, as long as one can integrate all these basis functions up to a certain degree n exactly, one can also integrate exactly any polynomial of degree less than or equal to n. Furthermore, it is useful to start by not being concerned with the values for the integration limits a and b. Instead, one can work on a convenient fixed interval, by choosing for example $a = -1$ and $b = 1$, then extend the results obtained on this interval to any arbitrary (but finite) interval through a change of variables. Since formulae developed this way should be used to eventually integrate arbitrary functions, consider an approximation of the form:

$$\int_{-1}^1 f(x)\,dx \approx \sum_{i=1}^N w_i f(x_i),$$

where now one can choose the points x_i and also the weights w_i such that the formula is exact for the largest possible class of polynomials, or equivalently for the largest possible set of basis functions. For example, if $N = 1$ one would have:

$$\int_{-1}^1 f(x)\,dx \approx w_1 f(x_1)$$

and there are two parameters that can be chosen, w_1 and x_1. One can choose them by asking that this formula produce an exact result for a corresponding number of basis functions, i.e. for the two basis functions $\{1, x\}$. By doing so, one is guaranteed to end up with a formula that will produce the exact result for any polynomial of degree one. Continuing along this line, for $N = 2$ the corresponding formula will be

$$\int_{-1}^1 f(x)\,dx \approx w_1 f(x_1) + w_2 f(x_2), \tag{9.5}$$

where there is no need to force w_1 and x_1 to have the same values as for the case $N = 1$; instead, these will be new values which, together with the values for w_2 and x_2, will allow one to use this formula for $N = 2$ to integrate exactly polynomials up to degree three.

Therefore, to find the coefficients, start by choosing $N = 1$ for example. Then the following two integrals have to produce exact results:

$$\int_{-1}^{1} f(x)\,dx = 2 = w_1 f(x_1), \quad \text{for } f(x) = 1,$$

$$\int_{-1}^{1} f(x)\,dx = 0 = w_1 f(x_1), \quad \text{for } f(x) = x.$$

From the first equality one gets $w_1 = 2$, as $f(x_1) = 1$, whatever the value of x_1 is. Using this in the second equality $2f(x_1) = 0$, where now $f(x_1) = x_1$, it follows that $x_1 = 0$. Therefore the Gauss formula for $N = 1$ has the form

$$\int_{-1}^{1} f(x)\,dx \approx 2f(0),$$

and this is guaranteed to be exact if $f(x)$ is a polynomial of degree one or less.

The same procedure is applied for the case $N = 2$; in this case the two coefficients and the corresponding integration points in equation (9.5) can be obtained by asking for an exact evaluation of the integral for an arbitrary cubic polynomial, or equivalently for the set of four basis functions $\{1, x, x^2, x^3\}$. Thereupon, one obtains the following system for the four parameters:

$$\int_{-1}^{1} f(x)\,dx = 2 = w_1 + w_2, \quad \text{for } f(x) = 1,$$

$$\int_{-1}^{1} f(x)\,dx = 0 = w_1 x_1 + w_2 x_2, \quad \text{for } f(x) = x,$$

$$\int_{-1}^{1} f(x)\,dx = \frac{2}{3} = w_1 x_1^2 + w_2 x_2^2, \quad \text{for } f(x) = x^2,$$

$$\int_{-1}^{1} f(x)\,dx = 0 = w_1 x_1^3 + w_2 x_2^3, \quad \text{for } f(x) = x^3.$$

Because the integration interval is symmetric about $x = 0$, one can assume $x_1 = -x_2$. From the first two equations, it then follows that $w_1 = w_2 = 1$. The fourth equation is then automatically satisfied under these conditions. Substituting these conditions into the third equation one then obtains:

$$2x_1^2 = \frac{2}{3} \quad \Rightarrow x_1 = -\sqrt{\frac{1}{3}} \approx -0.57735027 \text{ and } x_2 = \sqrt{\frac{1}{3}} \approx 0.57735027.$$

One of the difficulties encountered with this approach should, however, be apparent: the system of equations that determines the values of the integration parameters w_i and x_i is nonlinear for $N \geq 2$. While this system can be readily solved analytically for $N = 2$, as shown above, the solution for larger values of N is not straightforward. Table 9.1 lists the values of the Gauss points and weights up to $N = 6$. Results for the higher values of N can more readily be obtained in a different way that involves the use of the orthogonal Legendre polynomials, which is not elaborated upon here.

It is worth emphasizing that, in general, if $P(x)$ is a polynomial of degree less than or equal to $2N - 1$ then Gauss quadrature formulae will produce an exact result because there are $2N$ parameters in the formula (the coefficients w_i and the points x_i, for $i = 1, 2, \ldots, N$), the same number as the number of basis functions for a polynomial of degree $2N - 1$. In other words,

$$\int_{-1}^{1} P(x)\, dx = \sum_{i=1}^{N} w_i P(x_i),$$

for any polynomial $P(x) = a_0 + a_1 x + a_2 x^2 + \ldots + a_{2N-1} x^{2N-1}$.

TABLE 9.1: Gauss points and their associated weights.

N	Roots x_i	Coefficients w_i
1	0.00000000	2.00000000
2	0.57735027	1.00000000
	−0.57735027	1.00000000
3	0.77459667	0.55555556
	0.00000000	0.88888889
	−0.77459667	0.55555556
4	0.33998104	0.65214515
	0.86113631	0.34785485
	−0.33998104	0.65214515
	−0.86113631	0.34785485
5	0.90617985	0.23692689
	0.53846931	0.47862867
	0.00000000	0.56888889
	−0.53846931	0.47862867
	−0.90617985	0.23692689
6	0.93246951	0.17132449
	0.66120939	0.36076157
	0.23861919	0.46791393
	−0.23861919	0.46791393
	−0.66120939	0.36076157
	−0.93246951	0.17132449

To approximate the integral of a function $f(\xi)$ on a generic interval $[a, b]$ one can simply use a change of variables defined by a linear map,

$$\xi = \frac{b-a}{2}x + \frac{b+a}{2}, \quad -1 \leq x \leq 1 \quad \Longrightarrow \quad d\xi = \frac{b-a}{2}dx$$

and so the Gauss Quadrature rule on a general interval $[a, b]$ is given by

$$\int_a^b f(\xi)\, d\xi = \int_{-1}^1 f\left(\frac{b-a}{2}x + \frac{b+a}{2}\right)\frac{b-a}{2}dx$$

$$\approx \sum_{i=1}^n w_i \cdot f\left(\frac{b-a}{2}x_i + \frac{b+a}{2}\right) \cdot \frac{b-a}{2}. \tag{9.6}$$

9.3   Extrapolation Methods: Romberg Quadrature

The extrapolation technique discussed in section 7.5 for the central approximation to the first derivative can be used in conjunction with the trapezoidal method, for which the truncation error has the same form as the error for the central approximation to the first derivative. The resulting extrapolation method is known as Romberg quadrature and may be expected to improve approximations generated by the composite trapezoidal rule. In this case the quantity to be approximated is the integral $\int_a^b f(x)\, dx$. Its composite trapezoidal rule approximation is given by:

$$Q = \int_a^b f(x)\, dx = \frac{h}{2}\left[f(a) + 2\sum_{j=1}^{n-1} f(x_j) + f(b)\right] - \frac{(b-a)}{12}h^2 f''(\xi).$$

Notice that the first-order error term is absent in the composite trapezoidal approximation; the largest error term is $\mathcal{O}(h^2)$. Just like the case of central finite difference formulae, it can be shown that all odd-order error terms are in fact absent from the trapezoidal rule truncation error. Therefore, the first elimination to be performed in this case will again correspond to equation (7.13).

To use this extrapolation technique, the first step is to obtain approximations with a sequence of step sizes. These can be seen to correspond to a variable number of points n in the above formula, where n takes, in turn, the values $n = n_1, n_2, \ldots, n_k, \ldots$ with $n_1 = 1$, $n_2 = 2$, $n_3 = 4$ and in general $n_k = 2^{k-1}$. The step size for a mesh with $n = n_k$ patches is given by:

$$h_k = \frac{b-a}{n_k} = \frac{b-a}{2^{k-1}}$$

Therefore, on the k-th mesh, the composite trapezoidal rule approximation, which will be denoted by $R_{k,1}$, is given by:

$$\int_a^b f(x)\,dx = \underbrace{\frac{h_k}{2}\left[f(a) + 2\sum_{j=1}^{2^{k-1}-1} f(a+jh_k) + f(b)\right]}_{R_{k,1}} - \frac{(b-a)}{12}h_k^2 f''(\xi_k)$$

where

$$R_{1,1} = \frac{b-a}{2}[f(a) + f(b)],$$

$$R_{2,1} = \frac{1}{2}[R_{1,1} + h_1 f(a+h_2)],$$

$$R_{3,1} = \frac{1}{2}\left[R_{2,1} + h_2[f(a+h_3) + f(a+3h_3)]\right],$$

and, in general,

$$R_{k,1} = \frac{1}{2}\left[R_{k-1,1} + h_{k-1}\sum_{j=1}^{2^{k-2}} f\big(a+(2j-1)h_k\big)\right] \qquad \text{for } k = 2, 3, \ldots, m.$$

These expressions are just rearrangements for the sums that occur in the composite trapezoidal method for the particular sequence of step sizes created by halving the interval between successive points in the discretization. Their recursive formulation allows one to only compute function values at the additional points introduced with every new step reduction. Nevertheless, the sequence of approximations $R_{1,1}, R_{2,1}, \ldots$ converges with only second-order accuracy. By applying the extrapolation method explained above, one can speed up convergence considerably. For example, once $R_{1,1}$ and $R_{2,1}$ have been computed, an extrapolated value can be obtained as per equation (7.12):

$$R_{2,2} = \frac{4R_{2,1} - R_{1,1}}{3},$$

which may be expected to have eliminated the $\mathcal{O}(h^2)$ error term and thus have a truncation error of order $\mathcal{O}(h^4)$. Continuing in this manner, one computes the sequence of approximations:

$$R_{k,j} = \frac{4^{j-1}R_{k,j-1} - R_{k-1,j-1}}{4^{j-1} - 1}, \quad j = 2, 3, \ldots,$$

with the error in $R_{k,j}$ expected to be of order $\mathcal{O}(h^{2j})$. The values thus produced are usually arranged in a tabular format, where every new row represents an extra halving of the step size and each additional column an extra even power of h eliminated from the truncation error.

$R_{1,1}$

$R_{2,1}$ $R_{2,2}$

$R_{3,1}$ $R_{3,2}$ $R_{3,3}$

$R_{4,1}$ $R_{4,2}$ $R_{4,3}$ $R_{4,4}$

$\vdots$ $\vdots$ $\vdots$ $\vdots$ $\ddots$

$R_{n,1}$ $R_{n,2}$ $R_{n,3}$ $R_{n,4}$ $\cdots$ $R_{n,n}$

To see how Romberg integration works in detail, consider the quantity Q to be approximated to be the integral

$$I[0,2] = \int_0^2 xe^{-x^2}\,dx \approx 0.49084218055563290985.$$

With $f(x) = xe^{-x^2}$, the sequence of approximations proceeds as follows:

$$R_{1,1} = \frac{2-0}{2}[f(0) + f(2)] \approx 0.03663128;$$

$$R_{2,1} = \frac{1}{2}[R_{1,1} + 2f(1)] \approx 0.38619508;$$

$$R_{2,2} = \frac{4R_{2,1} - R_{1,1}}{3} \approx 0.50271635;$$

$$R_{3,1} = \frac{1}{2}\left[R_{2,1} + \frac{1}{2}\left(f(0.5) + f(1.5)\right)\right] \approx 0.46684715;$$

$$R_{3,2} = \frac{4R_{3,1} - R_{2,1}}{3} \approx 0.49373118;$$

$$R_{3,3} = \frac{16R_{3,2} - R_{2,2}}{15} \approx 0.49313217,$$

and so on. The following table shows the results obtained after another two subdivisions.

k	h	$R_{k,1}$	$R_{k,2}$	$R_{k,3}$	$R_{k,4}$	$R_{k,5}$
1	2	0.03663128				
2	1	0.38619508	0.50271635			
3	0.5	0.46684715	0.49373118	0.49313217		
4	0.25	0.48493688	0.49096679	0.49078249	0.49074520	
5	0.125	0.48937135	0.49084951	0.49084169	0.49084263	0.49084301

9.4 Higher-Dimensional Integrals

The methods presented so far can be immediately extended to those area and volume integrals that can be written as iterated integrals. To start with, consider the case of an area integral of the form:

$$I = \iint_D f(x,y)dA; \quad D = \{(x,y) \in \mathbb{R}^2 | a \leq x \leq b, c \leq y \leq d\},$$

where, in the simplest case, $f(x,y) = f_1(x) \cdot f_2(y)$. Then the area integral becomes the product of two one-dimensional integrals and can be written:

$$I = \iint_D f(x,y)dA = \int_a^b f_1(x)dx \cdot \int_c^d f_2(y)dy,$$

and similarly for a volume integral. A generalization of this method, sometimes referred to as the *parametrization method*, addresses domains of the form $D = \{(x,y) \in \mathbb{R}^2 | a \leq x \leq b, \gamma(x) \leq y \leq \delta(x)\}$. In this case, the area integral becomes:

$$I = \iint_D f(x,y)dA = \int_a^b \left[\int_{\gamma(x)}^{\delta(x)} f(x,y)dy \right] dx = \int_a^b g(x)dx,$$

where the inner integral, enclosed here in square brackets for clarity, has been denoted by $g(x)$. The task of approximating the area integral I is thus again reduced to evaluating two one-dimensional integrals. The extension to functions of three variables is straightforward. A numerical example showing how to perform this evaluation is given below.

9.4.1 Parametrization method example

The goal here is to evaluate the integral:

$$I = \iint_D (x + 2y)\, dA, \tag{9.7}$$

where D is the region in the xy-plane bounded by the parabolas $y = 2x^2$ and $y = 1 + x^2$, as shown in Figure 9.3. Notice that the intersection points of the two parabolas are $(-1, 2)$ and $(1, 2)$. The exact value of this integral is $I = 32/15$.

Let us first look into what it means to apply the parameterization technique to this integral. To that end, one needs to define the following auxiliary function:

$$g(x) = \int_{2x^2}^{1+x^2} (x + 2y)\, dy, \quad -1 \leq x \leq 1.$$

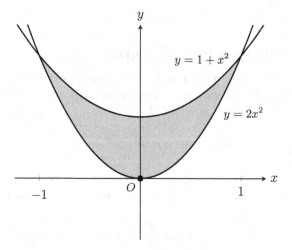

Figure 9.3: 2D Parameterization example.

Notice that, once a value of x is specified, the value of the function $g(x)$ is readily obtained by integrating with respect to y only. Also notice that one can choose among a variety of methods for evaluating both the outer and the inner integral. The chosen methods could be the same, but they don't need to. This example will use different methods for the outer x-integral and the inner y-integral.

The original area integral can be written, with our definition, as:

$$I = \int_{-1}^{1} g(x)\, dx.$$

Let us evaluate this x-integral using Gauss quadrature with three nodes. The value of the integral will be given by:

$$I \approx w_1 g(x_1) + w_2 g(x_2) + w_3 g(x_3), \tag{9.8}$$

where

$$g(x_1) = \int_{2x_1^2}^{1+x_1^2} (x_1 + 2y)\, dy, \tag{9.9}$$

$$g(x_2) = \int_{2x_2^2}^{1+x_2^2} (x_2 + 2y)\, dy, \tag{9.10}$$

$$g(x_3) = \int_{2x_3^2}^{1+x_3^2} (x_3 + 2y)\, dy. \tag{9.11}$$

$$\tag{9.12}$$

Here the points x_i and weights w_i come from table 9.1 for $N = 3$. That is, these values are:

x_1	=	0.77459667	$w_1 = 0.55555556$
x_2	=	0.00000000	$w_2 = 0.88888889$
x_2	=	−0.77459667	$w_3 = 0.55555556$

It should already be obvious that the function values $g(x_1)$, $g(x_2)$ and $g(x_3)$ are not immediately available. Instead, they need to be computed themselves by performing the inner integral. It is worth noting that:

$$g(x) = \int_{2x^2}^{1+x^2} (x + 2y)\, dy$$

$$= x \int_{2x^2}^{1+x^2} dy + 2 \int_{2x^2}^{1+x^2} y\, dy$$

$$= x\left[y\right]_{2x^2}^{1+x^2} + \left[y^2\right]_{2x^2}^{1+x^2}$$

$$= x[1 - x^2] + [(1 + x^2)^2 - (2x^2)^2]$$

$$= 1 + x + 2x^2 - x^3 - 3x^4.$$

Thus, in this example, there is an easy analytic way to compute $g(x)$. Alternately, one can also integrate numerically to find the needed values of $g(x_1)$, $g(x_2)$ and $g(x_3)$. If one chooses to perform the inner y-integral analytically, one obtains:

$g(x_1)$	=	0.8102	$w_1 g(x_1) = 0.794355$
$g(x_2)$	=	1	$w_2 g(x_2) = 0.888889$
$g(x_3)$	=	1.4298	$w_3 g(x_3) = 0.450090$

A simple computation of the resulting Gauss approximation in equation (9.8) leads to $I \approx 2.133333$. This is in fact the exact result, truncated here to six decimal places. This should not come as a surprise: one obtains the exact result because Gauss integration with three points is exact for a polynomial of order five or less, while the function $g(x)$ is a polynomial of degree four. Thus, the outer x-integral is guaranteed to produce an exact result. Since the exact analytical result was used for the inner y-integral, the overall result is itself exact.

9.5   Monte Carlo Integration

The Monte Carlo method[1] takes a very different, more statistical approach to the approximation of integrals. In its simplest form, it is based on the mean value theorem written for the antiderivative $F(x)$ of a function $f(x)$, written in the form:

$$\int_a^b f(x)dx = F(b) - F(a) = F'(\xi)(b - a) = f(\xi)(b - a),$$

where the value $\bar{f} = f(\xi)$ is naturally interpreted as the mean value of the function $f(x)$ on the interval $[a, b]$. From the opposite point of view, if the mean value of $f(x)$ is known or can be obtained through some other means, the integral can be approximated using the relationship above. The strong law of large numbers and the central limit theorem, important results in probability theory [17], guarantee that the following approximation:

$$\int_a^b f(x)dx = (b - a)\bar{f} \approx S_N = \frac{b - a}{N} \sum_{i=1}^N f(x_i),$$

where the N points are chosen randomly in the interval $[a, b]$ according to a uniform distribution, converges to the exact result as $N \to \infty$ with a mean error of the form:

$$\int_a^b f(x)dx = \frac{b - a}{N} \sum_{i=1}^N f(x_i) + \mathcal{O}(\frac{1}{\sqrt{N}}).$$

For integrals in one dimension this error behavior does not compare favorably with other methods. Looking at the trapezoidal method for example, if N is the number of function evaluations performed, then the spacing between the data points is on the order of $h = \mathcal{O}(1/N)$ so that the error for the trapezoidal method for the same amount of work (i.e. the same number of function evaluations) behaves as $\mathcal{O}(h^2) = \mathcal{O}(1/N^2)$. The comparison with Simpson's method is even worse. However, the error involved in the Monte-Carlo method only involves the number of points and is not related to any spacing; therefore the method compares ever more favorably with other methods as the dimensionality of the problem increases. That is so because in two dimensions say, the spacing $h = \Delta x = \Delta y$ is now of the order $\mathcal{O}(1/\sqrt{N})$ for N function evaluation (or grid) points, so the trapezoidal method for example will now be reduced to $\mathcal{O}(h^2) = \mathcal{O}(1/N)$, whereas the error in the Monte-Carlo method remains of

---

[1]The name comes from the fact that Monte Carlo, in the Principality of Monaco, is a very well-known gambling hub. U.S. readers, who may feel disadvantaged, relax: a subset of methods proudly showcases Las Vegas.

the same order with respect to N. It is also worth mentioning the fact that the expression for the Monte-Carlo error involves a statistical mean; that means that performing a Monte-Carlo simulation with N random function samples several times will not necessarily lead to the same error (unless, of course, the same sample points end up being reused).

The main tool necessary for performing Monte-Carlo quadrature is a uniformly-distributed random number generator, which MATLAB readily provides through the function **rand**. This function generates matrices of random numbers uniformly distributed in the interval $[0, 1]$ and details of its implementation and use can be accessed through its help page. To obtain x-locations that are uniformly distributed in $[a, b]$, the output from the function **rand** must be mapped accordingly. The script 9.2 shows the straightforward implementation of the method for computing the integral of the function $f(x) = x \cos(x)$ on the interval $[0, \pi/2]$. Remember that, unless the same random seed is used, the output from the script will be different between different runs due to the different x-locations for function evaluation.

```
MATLAB Script 9.2: 1D Monte-Carlo Method

clear all;        % clears all variables
f = @(x) x .* cos(x);
a = 0    ;        % left boundary
b = pi/2;         % right boundary
N = 100;
x = a + (b-a)*rand(1,N);
fbar = sum( f(x) ) / N;
integralMC = (b-a) * fbar      % Monte-Carlo
integralNC = integral(f, a, b) % MATLAB (Newton-Cotes)
% sample output
integralMC =   0.58478
integralNC =   0.57080
```

A great advantage of Monte Carlo methods is their straightforward extension to multiple dimensions in the case when the domain has a simple box shape. Suppose for example that one wants to approximate the following integral over a rectangle:

$$I = \iint_R f(x, y) dA; \quad R = \{(x, y) \in \mathbb{R}^2 | a \le x \le b, c \le y \le d\}.$$

The mean value theorem tells us that in this case the approximation will be:

$$I = \iint_R f(x, y) dA \approx \frac{\mathbf{A}(R)}{N} \sum_{i=1}^{N} f(x_i, y_i),$$

where $\mathbf{A}(R)$ is the area of the rectangle R. Taking $f(x, y) = x^2 + xy$ for an example, the sample code 9.3 computes its integral by first generating N points with coordinates (x_i, y_i) such that x_i is uniformly distributed in $[a = 1, b = 3]$ and y_i in $[c = -1, d = 2]$:

MATLAB Script 9.3: 2D Monte-Carlo Method

```
clear all;        % clears all variables
f = @(x,y) x.^2 + x .*y;
a = 1  ;          % left x boundary
b = 3;            % right x boundary
c = -1;           % left y boundary
d = 2;            % right y boundary
N = 100;
x = a + (b-a)*rand(1,N);
y = c + (d-c)*rand(1,N);
fbar = sum( f(x,y) ) / N;
integralMC = (b-a) * (d-c) *  fbar       % Monte-Carlo
% sample output
integralMC =   32.847
```

While not immediately obvious, when the region of integration cannot be expressed as the simple Cartesian product of one-dimensional intervals, the method can still be applied in a number of ways. Let us denote the region of integration by D in this case. One way to perform Monte Carlo on D is to create a random number generator that can generate uniformly distributed points in D. This can be done in some cases, but will not be explored here. The other alternative is to embed the actual domain of integration D in a box (a rectangle in two dimensions) R, and define an extension function over the embedding domain R. In this case, the integral is computed as:

$$I = \iint_D f(x,y)dA = I = \iint_R \tilde{f}(x,y)dA \approx \frac{\mathbf{A}(R)}{N} \sum_{i=1}^{N} \tilde{f}(x_i, y_i),$$

where the function $\tilde{f}(x,y)$ is defined as follows:

$$\tilde{f}(x,y) = \begin{cases} f(x,y), & (x,y) \in D \\ 0, & (x,y) \in R \setminus D \end{cases}$$

A classical problem that can be easily addressed in this way and at the same time helps us clarify the concepts involved is estimating the value of the constant π; it can be thought of as estimating the area of a disk by throwing darts. Consider the domain D to be the disk of unit radius and center at the origin in $\mathbb{R}^2$, that is $D = \{(x,y) \in \mathbb{R}^2 | x^2 + y^2 \leq 1\}$, and let the embedding rectangle R, in this case actually a square, be defined by $R = \{(x,y) \in \mathbb{R}^2 | -1 \leq x, y \leq 1\}$. The area of the disk D is given by:

$$\mathbf{A}(D) = \iint_D f(x,y)dA = \pi \cdot 1^2 = \pi; \ \ \mathbf{with} f(x,y) = 1,$$

so that by defining

$$\tilde{f}(x,y) = \begin{cases} 1, & (x,y) \in D \\ ,0, & (x,y) \in R \setminus D, \end{cases}$$

one can approximate the area of the disk, and hence the value of π, by performing Monte Carlo quadrature on the square R instead. The script 9.4 implements this idea.

```matlab
MATLAB Script 9.4: 2D Monte-Carlo: Estimating π

clear all;        % clears all variables
a = -1 ;          % left x boundary
b = 1;        % right x boundary
c = -1;           % left y boundary
d = 1;        % right y boundary
N = 10000;
x = a + (b-a)*rand(1,N);
y = c + (d-c)*rand(1,N);
sum = 0;
for i = 1:N
    sum = sum + fExt( x(i), y(i) );
end
fbar = sum / N ;
integralMC = (b-a) * (d-c) *  fbar       % Monte-Carlo
% The extended function (can be saved in fExt.m)
function [val] = fExt ( x, y )
    val = 0;
    if x^2 + y^2 <= 1
        val = 1;
    end
end
% sample output
integralMC =  3.1348
```

9.6 MATLAB® Built-in Functions

The MATLAB function `integral` takes as arguments a function and the limits of the interval for integration and returns the estimated value of the integral. The function is somewhat more complicated than the procedures described here, in that the integration is adaptive: it uses a numerical estimate for the error in the integration process, which it tries to keep under a certain tolerance. The function can also evaluate some improper integrals; to calculate the integral $\int_1^\infty f(x)dx$ for example, the corresponding call is `integral( f, 0, Inf)`. The function `quad2d` can be used to evaluate integrals over a planar region as described in section 9.4 and is similar to `integral2` in terms of its input parameters. For example, `integral2` takes as arguments a function of two variables x and y, fixed limits for the domain of integration in the x-direction and either constant or x-dependent limits in the y-direction. Three-dimensional integrals can similarly be evaluated with `integral3`.

Finally, MATLAB offers a Gauss-type adaptive quadrature rule through the function `quadgk`.

9.7　Exercises

Ex. 1 Calculate the value of

$$\int_0^{2\pi} x \sin^2 x \, dx$$

using both the trapezoidal rule and Simpson's rule. For a fair comparison, keep the same number of function evaluations, in this case using five equally-spaced points $\{x_0, x_1, x_2, x_3, x_4\}$. Do this calculation by hand, truncating to four decimal places throughout.

Ex. 2 Write a MATLAB function that uses the composite trapezoidal rule to evaluate the same integral in the above example. Your code should be more general, in that it should accept the number of patches n (which you can think of as being equivalent to specifying the number of function evaluations, $n + 1$) as an input. Compare the result with your hand calculation.

Ex. 3 Use the function in the preceding exercise to study the convergence of the composite trapezoidal rule for the integral

$$I = \int_0^2 \exp(x) dx.$$

Start with $n = 2$ and perform a total of three approximations, decreasing the distance h between the evaluation points by a factor of two for each successive approximation. Plot the absolute error as a function of the step size h using a semi-logarithmic plot, with the logarithmic scale on the vertical axis.

Ex. 4 Repeat the calculation in the previous exercise using a code of your own that implements Simpson's method. This time, plot the errors for both the composite trapezoidal rule and composite Simpson's rule on the same semi-logarithmic plot.

Ex. 5 Again repeat the same calculation, this time using Romberg quadrature up to the $R_{3,3}$ approximation for the same integral. Do this calculation by hand, rounding off to five decimal places. Note that you can reuse results from your solution to **Ex. 3**. Compute the error for this approximation, $E_{3,3} = |R_{3,3} - (e^2 - 1)|$, and compare it with the error for the composite trapezoidal rule on the same grid.

Ex. 6 In some particular situations, Romberg quadrature seems to behave contrary to expectations. Study the behavior of the composite trapezoidal rule and the Romberg extrapolation for the integral

$$I = \int_0^\pi x \sin^2(x)\,dx = \frac{\pi^2}{4}.$$

Create the Romberg approximations up to $R_{3,3}$. Do this calculation by hand, rounding off to five decimal places. Compare the value obtained for $R_{3,3}$ with the value obtained from the composite trapezoidal rule, $R_{3,1}$. Can you explain the result in this case?

Ex. 7 Write a MATLAB function that implements Romberg quadrature to compute the $R_{m,m}$ approximation for m a positive integer specified by the user. Together with m, the inputs should include the function to be integrated and the interval of integration. Test your function against the result from the previous problems.

Ex. 8 Evaluate the integral (9.7) using Gauss quadrature in the x-direction and Simpson's rule in the y-direction using a MATLAB implementation of your own. Your code should allow for up to three Gauss points and any user-specified number of panels for the Simpson integral.

Ex. 9 Modify the MATLAB script 9.4 to again evaluate the integral (9.7), this time using the Monte-Carlo method. Chose the rectangle $R = \{(x,y) : [-1,1] \times [0,2]\}$ to cover the area of D. Notice that the area of D can be computed by $\int_{-1}^{1}(1 - x^2)\,dx = 4/3$ and let $\gamma = \frac{\text{area of } D}{\text{area of } R} = \frac{1}{3} < \frac{1}{2}$. The consequence of a small γ is that a large number of samples is required to obtain an accurate approximation. Compare your results for M samples, with $M = 10,000$ and $M = 10,000,000,000$.

Ex. 10 Write a code that integrates the function $f(x,y,z) = 0.7(x^2 + y^2 + z^2)$ over the unit sphere $S = \{(x,y,z) \mid x^2 + y^2 + z^2 \le 1\}$ using the Monte-Carlo method in three dimensions. Run the code with a sample of $M = 10^2$ points; do this a total of $T = 100$ times and compute the average of these one hundred results, A_{MT}. Then calculate the absolute error E_{MT} in this average by comparing it with the exact value of the integral, which can be easily calculated analytically. Plot the error E_{MT} versus M for a fixed value of $T = 100$ and $M = 10^2$, $M = 10^4$ and $M = 10^6$. The error is naturally defined here as the absolute value of the difference between the exact value and the average E_{MT}.

Ex. 11 Write a MATLAB code that uses the parametrization method to calculate the value of the area integral

$$I = \int_{-1}^{1} \int_{x^2}^{4-x^2} \left(x^2 y + e^{-(x^2+y^2)} \right) dy\,dx.$$

Use Gauss quadrature in the x-direction (the outer integral) and either Simpson's rule or Gauss quadrature, or both, in the y-direction (the inner integral). For the Gauss integrals, the code should allow you to choose up to five integration points. Similarly, if you choose Simpson's method for the inner integral, you should be able to choose an arbitrary number of panels, K; the number of patches will then be $n = 2K$ with the evaluation points numbered $y_0, y_1, y_2, \ldots, y_n$. Run the code with $N = 2$ and $K = 2$, then steadily increase the values of both parameters N and K. Do the results you obtain show a converging trend?

Chapter 10

Numerical Solution of Differential Equations

Ordinary differential equations (ODEs) and their more complicated relatives, partial differential equations (PDEs), are encountered in most areas of science and engineering. They allow us to create, starting from physical laws such as conservation of mass and energy, models that can be used as prediction tools. For this reason, they can be encountered in fields as far apart as population dynamics, weather forecasting and black-hole collision modeling. Since no previous exposure to the subject matter is expected, this chapter starts with a general presentation of both the nomenclature and the way some of the more common differential equation models are established. It then moves on to establishing numerical methods for their solution. For a more detailed fundamental treatment the reader is encouraged to consult a specialized work, such as the classical text by Boyce and DiPrima [1].

10.1   First-Order Models

A first-order differential equation is a statement of the form

$$\frac{du}{dt} = f(u, t) \tag{10.1}$$

about the derivative of a function $u(t)$, which is known as the *dependent variable* and constitutes the unknown to be solved for. The dependent variable is a function of the *independent variable*, denoted here by t. For the models discussed below the independent variable usually has the physical meaning of time, although this may not always the case in general. To solve the equation on an interval (a, b) means to find a function $\nu(t)$ which, when substituted for $u(t)$ in the equation, transforms the equation into an identity, i.e. the equality is satisfied for all values $t \in (a, b)$. Notice that, at least in principle, finding an expression for the dependent variable involves an integral, so the solution is defined only up to a constant. Nevertheless, an explicit solution for the dependent variable as a function of the independent variable is not always possible. When an explicit solution is possible, such a solution containing an unspecified constant is known as the *general solution* if all possible solutions are obtained by simply varying the value of this constant. To further specify the solution, knowledge of the dependent variable $u(t)$ at some value $t = t_0$ is necessary. Such knowledge is usually obtained from measurement and takes the form of an *initial condition* $u(t_0) = u_0$. A first-order differential equation with a provided initial condition forms an *initial value problem*. Furthermore, we assume that the initial value problems presented below do have a unique solution. Although a thorough mathematical analysis of the existence and uniqueness of solutions of differential equations is beyond an elementary treatment of the subject, the reader is again encouraged to consult [1].

10.1.1   Classical mechanics

Newton's second law of motion leads quite naturally to such first-order differential equations. Consider, for the simplest possible example, the motion of a compact body of mass m, in free fall under the effect of gravity. The law of motion states that the force acting on the body is $\vec{F} = m\vec{a} = m\vec{g}$, where $\vec{a} = d\vec{v}/dt$ is the acceleration exhibited by the body, $\vec{v}(t)$ its velocity and $m\vec{g}$ is the force due to the object being in the gravity field of the Earth. Upon projection on the vertical direction (with positive sign usually chosen towards the surface of the Earth) the vectorial equation of motion becomes a differential equation for the velocity magnitude $v(t)$, $dv/dt = g$. This can be solved immediately, in what amounts to a simple calculus exercise, to yield $v(t) = v_0 + gt$, with v_0 the initial velocity at time $t = 0$ and g the gravitational constant.

The difficulty that most differential equations pose in their solutions is obscured in this first example, but becomes obvious when considering a slightly nore complicated scenario: fall with air drag. Specifically, consider that the motion of the body through the air is subject to a drag force, due to the ensuing friction, that always opposes motion and is proportional to the square of the velocity magnitude. In this case, the projection of the equation of motion onto the vertical direction yields:

$$m\frac{dv}{dt} = mg - \alpha v^2, \quad v(t = 0) = v_0,$$

where $\alpha > 0$ is the proportionality constant; the minus sign on the right accounts for drag direction. A naive attempt to solve for $v(t)$ from $dv/dt = g - (\alpha/m)v^2$ will fail because the unknown function $v(t)$ appears on the right-hand side of the equation. That is, the integral $\int (g - \alpha v^2/m)dt$ cannot be evaluated in the absence of an expression for $v(t)$ (hence also for $v^2(t)$) as an explicit function of the independent variable t.

10.1.2 Radioactive decay

Consider a large sample of material that contains unstable, radioactive nuclei that are subject to decay through emission of elementary particles. The decay of one particular unstable nucleus is totally random and can happen at any moment in time, with equal likelihood. If the number of unstable nuclei in the sample at some particular time is $N(t)$, the number of nuclei ΔN that may be expected to decay within a certain time interval Δt is proportional to $N(t)$, $\Delta N = \lambda N(t)$, $\lambda > 0$. The rate of change in the number of unstable nuclei per unit time is thus $-\Delta N/\Delta t$, where the minus sign is needed because $N(t)$ decreases. Since a decay can occur within any time interval, however small, one can take the limit as $\Delta t \to 0$ to obtain the following radioactive decay initial value problem

$$\frac{dN}{dt} = -\lambda N(t), \quad N(t = 0) = N_0. \tag{10.2}$$

This rather simple equation, together with its counterpart governing growth to be seen in the next subsection, has a big role to play in numerical methods. The analytical solution is $N(t) = N_0 e^{-\lambda t}$, which can checked as follows: the equation is satisfied, since

$$\frac{dN(t)}{dt} = \frac{d(N_0 e^{-\lambda t})}{dt} = N_0(-\lambda)e^{-\lambda t} = -\lambda(N_0 e^{-\lambda t}) = -\lambda N(t),$$

while the initial condition also trivially holds because $N(t = 0) = N_0 e^0 = N_0$.

10.1.3 Population dynamics models

Suppose a very large population $p(t)$ of organisms that reproduce by subdivision is observed in a laboratory experiment. The organisms are provided with ideal conditions and the observation time is very long as compared to the time it takes for them to subdivide. If every individual subdivides into two identical copies of itself within a time Δt, the net increase in the population over this time interval is $\Delta p = p(t)$; that is, just as many new individuals have sprung up through subdivision as there were at the beginning of the interval. Since the observation happens on a very large time scale, it is again appropriate to consider the instantaneous rate of change $\lim_{\Delta t \to 0} \Delta p/Deltat$, which leads to the following initial value problem:

$$\frac{dp}{dt} = rp, \quad p(t = 0) = p_0. \tag{10.3}$$

Here the constant r, which has as unit of measure the reciprocal of the unit for time, has been introduced to make the two quantities on the left and right of the equal sign dimensionally consistent. In the scenario discussed so far, the value of r would be one; an immediate generalization can be considered by allowing $r > 0$ to be the *rate of growth* of the population, dictated by the speed at which some given population produces offspring. The model can then be used for a variety of species, as long as they encounter favorable reproductive conditions.

It is easy to see that the solution of equation (10.3) is $p(t) = p_0 e^{rt}$. That is, the population exhibits exponential growth for $r > 0$. It is helpful to consider here a useful convention that is rather the rule in the realm of differential equations. Notice that the constants that have been introduced in all the previous models have been all taken to be positive. This has a very good reason: compare the equation for exponential decay with the one for exponential growth. If λ in equation (10.2) were instead allowed to be negative, that equation would become $dN/dt = \lambda N$, since the right-hand side must be negative. For someone looking at this latter equation, there would be no immediate way to say whether the equation governs growth or decay; except for λ replacing the constant r, the equation would be similar to equation (10.3). In general, keeping the necessary constants positive helps us identify the role of different terms on the right-hand side: a positive sign on a term will indicate contribution to growth, while a minus sign will indicate decay.

Obviously, no population will be able to sustain a constant rate of growth over an indefinite period of time because of scarcity of resources. A model that adjusts the effective rate of growth so that it decays linearly while the population increases is provided by the *logistic equation*. The initial value problem in this case becomes:

$$\begin{cases} \frac{dp}{dt} = r\left(1 - \frac{p}{K}\right)p, \\ p(t = 0) = p_0, \end{cases} \tag{10.4}$$

where the constant r is known as the intrinsic growth rate.

All the above models involve only one stand-alone equation and can, in fact, be solved analytically [1]. Numerical methods become important when analytical solutions are not available. Probably the most important such instances arise when dealing with systems of equations. A model that involves two species and is simple to understand in its derivation is the *predator-prey population model*, also known as the *Lotka-Volterra model*. It leads to a system of two coupled first-order differential equations. Consider the population of a species, $x(t)$, that is preyed upon by a predator population $y(t)$. In the absence of the predator, the prey population is expected to grow exponentially, with growth rate a; conversely, in the absence of the prey population they feed on, the predator dies off exponentially, with a decay rate c. On the other hand, the prey is killed at a rate proportional to the number of encounters of members of the two species, which can be considered to be roughly proportional to the product xy; the predators are fed because of these encounters instead, hence they contribute to their growth. The system of equations that takes into account these considerations, together with the associated initial conditions, is therefore:

$$\frac{dx}{dt} = ax - \alpha xy; \quad x(t = 0) = x_0,$$

$$\frac{dy}{dt} = -cy + \gamma xy; \quad y(t = 0) = y_0.$$

Notice that these equations must be solved simultaneously, since the equation for $x(t)$ has $y(t)$ on the right-hand side and vice versa. The product term xy makes these equations nonlinear. Since there is no general method to solve such a nonlinear system analytically, numerical methods are of great importance.

10.1.4 Systems of first-order equations

The predator-prey equations represent a specific case of a first-order system of equations. Such systems are very important in practice, hence it's worth looking into the general case. A system of n equations for n dependent variables $x_1(t), x_2(t), \ldots, x_n(t)$ has the form

$$\frac{dx_1}{dt} = f_1(x_1, x_2, \ldots, x_n, t),$$

$$\frac{dx_2}{dt} = f_2(x_1, x_2, \ldots, x_n, t),$$

$$\vdots$$

$$\frac{dx_n}{dt} = f_n(x_1, x_2, \ldots, x_n, t),$$

and needs to be supplemented with n initial conditions of the form $x_1(t_0) = x_{10}$, $x_2(t_0) = x_{20}$ and so on; the initial time t_0 is usually taken to be zero. As seen before, a more convenient way to write the system is obtained by switching to vectorial quantities. The system becomes

$$\frac{d\vec{x}}{dt} = \vec{f}(\vec{x}, t),$$

where the vector of unknowns is defined to be $\vec{x}(t) = [x_1(t), x_2(t), \ldots, x_n(t)]^T$ and the right-hand side is now a column vector of functions, $\vec{f} = [f_1(\vec{x}, t), f_2(\vec{x}, t), \ldots, f_n(\vec{x}, t)]^T$.

As usual, the dependent variables will be denoted, for convenience, by either $\vec{x} = [x, y, z, \ldots]^T$ or $\vec{x} = [u, v, w, \ldots]^T$ when $n \leq 3$. In the most general case the functions appearing on the right-hand side are nonlinear in all their variables. The system is a first-order linear system if these functions are instead linear in all the dependent variables. In this case the system takes the form

$$\frac{dx_1}{dt} = a_{11}(t)x_1 + a_{12}(t)x_2 + \ldots + a_{1n}(t)x_n + g_1(t),$$

$$\frac{dx_2}{dt} = a_{21}(t)x_1 + a_{22}(t)x_2 + \ldots + a_{2n}(t)x_n + g_2(t),$$

$$\vdots$$

$$\frac{dx_n}{dt} = a_{n1}(t)x_1 + a_{n2}(t)x_2 + \ldots + a_{nn}(t)x_n + g_n(t).$$

If the functions $g_i(t)$ are all zero, the system is a *homogeneous system*. Moreover, when the coefficients of the dependent variables are constant, so that the independent variable t does not appear explicitly on the left-hand side, the system is known as a constant-coefficient homogeneous system. In this latter case, the system can be most conveniently written in the matrix form $d\vec{x}/dt = A\vec{x}(t)$.

10.2   Second-Order Models

Second-order ordinary differential equations are extremely valuable, in particular as models of mechanical and electrical vibrations. Only two such models will be briefly touched upon here; the user is encouraged to study a more in-depth treatment such as [1] for more details.

The motion of a body of mass m attached to an elastic spring of stiffness k and subject to energy dissipation characterized by a damping constant γ under the effect of an external excitation $F(t)$ is described by the second-order equation

$$m\frac{d^2x}{dt^2} + \gamma\frac{dx}{dt} + kx = F(t).$$

This equation is just a statement of Newton's second law of motion. When used with proper initial conditions, usually specified in the form $x(0) = x_0$ and $dx/dt(0) = v_0$ where x_0 and v_0 are the initial position and velocity of the body, respectively, it constitutes a second-order initial value problem. The parameters involved are again all considered positive in a physically realistic scenario, with $\gamma = 0$ allowed to describe the idealized case of a non-dissipative system. With an appropriate interpretation of the quantities involved, the same equation describes the oscillations of a simple electrical circuit.

The motion of a pendulum that consists of a blob of mass m hanging from a string of constant length L which is allowed to do small oscillations of angle θ in a fixed plane around the vertical position $\theta = 0$ is described by:

$$\frac{d^2\theta}{dt^2} + \frac{g}{L}\sin(\theta) = 0,$$

an equation that is again just an expression of Newton's second law in the tangential direction. In this direction, the tangential component of the force of gravity has magnitude

$$\left| \vec{F}_\theta \right| = mg \sin(\theta),$$

but acts in direction opposite to increasing θ, while the velocity component is $v(t) = d(L\theta)/dt$, so that the acceleration $a(t) = d^2(L\theta)/dt^2$. The equation considers that no other forces are present; in particular there is no energy loss due to friction.

10.3 Basic Numerical Methods

10.3.1 Euler's method

The *forward Euler method* (abbreviated FEM henceforth) may have been first encountered in an ODE class, or even a calculus class. The idea behind it is relatively straightforward. Consider a discretization of the time axis by points $t_i = t_0 + i\Delta t$, for $i = 0, 1, 2, \ldots$. For the generic equation (10.1) associated with a given initial condition of the form $u(t_0) = u_0$, discretize the derivative on the left-hand side using the forward first derivative approximation to obtain:

$$\frac{u(t_{i+1}) - u(t_i)}{\Delta t} \approx f(u_i, t_i), \quad i = 0, 1, 2, \ldots. \tag{10.5}$$

Notice that this is a forward approximation in the sense that the right-hand side is evaluated at the base point t_i, while the left-hand side approximation of the derivative uses the point $t_{i+1} > t_i$. From this equation one can express $u(t_{i+1})$, and thus advance the solution to the next time step:

$$u(t_{i+1}) \approx u(t_i) + \Delta t \cdot f(u_i, t_i), \quad i = 0, 1, 2, \ldots. \tag{10.6}$$

Also notice that all terms on the right-hand side are known. Because this is an approximation, it is useful to henceforth denote by u_i the numerical approximation of $u(t_i)$. With this notation, the FEM for a single time step from t_i to t_{i+1} can be written:

$$u_{i+1} = u_i + \Delta t \cdot f(u_i, t_i); \quad u_{i+1} \approx u(t_{i+1}). \tag{10.7}$$

One can also take a different point of view, easier to extend when searching for more accurate methods, and derive the FEM from an integral perspective. Consider a differential equation of the form $\frac{du}{dt} = f(u, t)$ and integrate formally in time to get:

$$u(t_{i+1}) = u(t_i) + \int_{t_i}^{t_{i+1}} f(u, t) dt, \quad i = 0, 1, 2, \ldots. \tag{10.8}$$

Since the unknown function $u(t)$ occurs in the integral, one cannot proceed unless some mechanism is used to approximate the integral on the right-hand side of equation (10.8). This can be done, however, with the numerical quadrature methods in Chapter 9.

Euler's method uses the simplest approximation for the integral: the function $f(u,t)$ is considered constant and equal to the value at the left endpoint t_i:

$$\int_{t_i}^{t_{i+1}} f(u,t)dt \approx \int_{t_i}^{t_{i+1}} f(u_i,t_i)dt = f(u_i,t_i) \cdot (t_{i+1} - t_i) = \Delta t \cdot f(u_i,t_i).$$

Hence, with this approximation, we unsurprisingly obtain again:

$$u(t_{i+1}) \approx u(t_i) + \Delta t \cdot f(u_i,t_i) \text{ or } u_{i+1} = u_i + \Delta t \cdot f(u_i,t_i).$$

10.3.2 Improving on Euler's method

For an improved method, one can use a more accurate approximation for the integrand in equation (10.8). For a first example, consider what happens when the integrand is approximated by an average over its values at the two endpoints. This amounts to assigning the following value for the integrand over the time interval $[t_i, t_{i+1}]$:

$$f(u,t) \approx \frac{f(u_i,t_i) + f(u_{i+1},t_{i+1})}{2},$$

which upon performing the integral amounts to:

$$u_{i+1} = u_i + \frac{\Delta t}{2}\left[f(u_i,t_i) + f(u_{i+1},t_{i+1})\right]. \tag{10.9}$$

This would be a valid formula in its own right, but upon inspection it does rise a problem: the value u_{i+1} on the left-hand side is the unknown one is trying to solve for and it also appears in the last term on the right-hand side. Solving for it requires therefore the solution of a nonlinear equation. Methods that take this approach are known as *implicit methods*, as opposed to the FEM, which is an *explicit method*.

A way to construct an explicit method out of the relationship (10.9) can be devised as follows. Let a completely explicit, predicted value for u_{i+1} be given by:

$$\tilde{u}_{i+1} = u_i + \Delta t \cdot f(u_i,t_i). \tag{10.10}$$

Then one can use this in place of the original u_1 on the right-hand side and consequently compute u_1 from:

$$u_{i+1} = u_i + \frac{\Delta t}{2}\left[f(u_i,t_i) + f(\tilde{u}_{i+1},t_{i+1})\right]. \tag{10.11}$$

Relationships (10.10) and (10.11) constitute a first example of a family of predictor-corrector methods known as *Runge-Kutta methods*. This particular one is known as Heun's method. It is possible to show, using more complicated Taylor series arguments, that the error for Heun's method is of order $\mathcal{O}(\Delta t^2)$ over a fixed time interval $[t_0, t_n]$, which represents a one order of accuracy improvement over the FEM[1]. Note that Heun's method can be thought of as an implementation of the trapezoidal rule for integral evaluation.

10.3.3 The classical Runge-Kutta method

As we already know from the chapter on numerical integration, Simpson's Rule is a more accurate approximation for integrals than the trapezoidal rule. Let's again see what happens if one tries to implement Simpson's rule for the evaluation of the integral in (10.8).

To this end, one first needs to divide the interval, of length Δt, from t_i to t_{i+1} into two subintervals (patches) of length $\frac{\Delta t}{2}$. Let $t^* = \frac{t_i + t_{i+1}}{2}$ be the midpoint of the interval from t_i to t_{i+1}, so that $u^* \approx u(t^*)$. Then Simpson's rule can be written:

$$\int_{t_i}^{t_{i+1}} f(u, t)dt \approx \frac{1}{3}\left(\frac{\Delta t}{2}\right)[f(u_i, t_i) + 4f(u^*, t^*) + f(u_{i+1}, t_{i+1})].$$

The same problem as before is apparent: the right-hand side terms cannot be explicitly evaluated. Proceeding as for Heun's method, one could however create predicted estimates for both u^* and u_{i+1}, both terms that appear on the right-hand side. A first estimate of u^* can be obtained from the straightforward application of the FEM:

$$\tilde{u}^* = u_i + \left(\frac{\Delta t}{2}\right)f(u_i, t_i),$$

where a step size $\frac{\Delta t}{2}$ needs to be used to advance from t_i to t^*. One can also consider a possible correction for this estimate defined by:

$$u^* = u_i + \left(\frac{\Delta t}{2}\right)f(\tilde{u}^*, t_i + \frac{\Delta t}{2}),$$

which clearly amounts to looking back from t^* to t_0 (as opposed to looking forward from t_0 to t^*, as the FEM does):

$$\tilde{u}_{i+1} = u^* + \left(\frac{\Delta t}{2}\right)f(u^*, t_i + \frac{\Delta t}{2}).$$

Finally, one uses all these preliminary, predicted values in Simpson's Rule by splitting the middle term in the sum on the right-hand side as follows:

$$u_{i+1} = u_i + \frac{\Delta t}{6}[f(u_i, t_i) + 2f(\tilde{u}^*, t^*) + 2f(u^*, t^*) + f(\tilde{u}_{i+1}, t_{i+1})],$$

---

[1]We will discuss the meaning of order of accuracy in Section 10.4

where $t^* = t_i + \Delta t/2$.

This method can be written in a more convenient, compact way if one lets $h = \Delta t$. Then, given (u_i, t_i), four stages need to be performed before the solution is advanced one time step:

$$K_1 = h \cdot f(u_i, t_i),$$

$$K_2 = h \cdot f(u_i + \frac{K_1}{2}, t_i + \frac{h}{2}),$$

$$K_3 = h \cdot f(u_i + \frac{K_2}{2}, t_i + \frac{h}{2}),$$

$$K_4 = h \cdot f(u_i + K_3, t_i + h),$$

$$u_{i+1} = u_i + \frac{1}{6}(K_1 + 2K_2 + 2K_3 + K_4).$$

This method is known as the classical, four-stage Runge-Kutta Method. It is usually denoted by RK4 since it is fourth-order accurate, $\mathcal{O}(\Delta t^4)$, over a fixed-size time interval.

10.4   Global Error and the Order of Accuracy

Consider the function

$$u(t) = e^{-4t} - t + 2, \tag{10.12}$$

where $u(0) = 3$. A simple calculation shows that:

$$u' + 4u = -4e^{-4t} - 1 + 4(e^{-4t} - t + 2) = 7 - 4t,$$

which suggests that the function u is the solution of the initial-value problem

$$u' = -4u + (7 - 4t)$$
$$u(0) = 3. \tag{10.13}$$

Suppose one solves the initial-value problem (10.13) by using the FEM (10.7) for $0 \leq t \leq 0.5$. In doing so, a series of step sizes, $\Delta t = 0.1, 0.05, 0.025, 0.0125$ and 0.00625 are employed. The total number of iterative steps corresponding to these step sizes amounts to $N = 5, 10, 20, 40$ and 80, respectively. Figure 10.1 plots the FEM solutions against the exact solution for $0 \leq t \leq 0.5$.

Denoting the value of the numerical solution at the final step by u_N, we define the *global error* between the computed solution and the exact solution at $t = 0.5$ as:

$$E_N = u(0.5) - u_N. \tag{10.14}$$

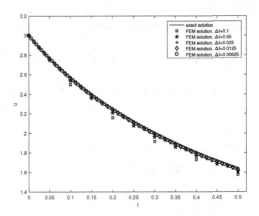

Figure 10.1: Computed FEM solutions to the initial-value problem (10.13), with step sizes $\Delta t = 0.1, 0.05, 0.025, 0.0125$ and 0.0625, respectively. The computed solutions are plotted against the exact solution for $0 \leq t \leq 0.5$.

Table 10.1 shows that when the step size is reduced by a factor of two, the global error is also roughly reduced by a factor of two, especially when Δt is small, as should be expected.

TABLE 10.1: Grid refinement study. The error $E_N = u(0.5) - u_N$ is the difference between the exact solution and the computed solution at $t = 0.5$; also shown is the ratio between the errors of two successive approximations, ratio $= \frac{(E_N)_{\Delta t}}{(E_N)_{\Delta t/2}}$.

	$\Delta t = 0.1$	$\Delta t = 0.05$	$\Delta t = 0.025$	$\Delta t = 0.0125$	$\Delta t = 0.00625$
E_N	0.0576	0.0280	0.0138	0.0068	0.0034
ratio		2.06	2.03	2.03	2.00

Let us now investigate the relationship between the step size and the global error of the FEM at the nth step, where $1 \leq n \leq N$. Note that the initial-value problem (10.13) belongs to a class of scalar linear initial-value problems of the form

$$u' = \lambda u + g(t)$$
$$u(t_0) = u_0. \tag{10.15}$$

When applying the FEM to (10.15), the numerical update relationship becomes:

$$u_{n+1} = u_n + \Delta t(\lambda u_n + g(t_n)), \tag{10.16}$$

where $t_n = t_0 + n\Delta t$ and $t_{n+1} - t_n = \Delta t$. Let $u(t_n)$ be the solution of (10.15) at t_n. Recall the Taylor's series expansion for $u(t_{n+1})$ with respect to t_n:

$$u(t_{n+1}) = u(t_n) + \Delta t u'(t_n) + \frac{1}{2}(\Delta t)^2 u''(t_n) + \mathcal{O}((\Delta t)^3). \tag{10.17}$$

By subtracting u_{n+1} in (10.16) from $u(t_{n+1})$ while applying (10.17) and (10.15) at t_n, one gets:

$$
\begin{aligned}
u(t_{n+1}) - u_{n+1} =\, & u(t_n) + \Delta t u'(t_n) + \frac{1}{2}(\Delta t)^2 u''(t_n) + \mathcal{O}((\Delta t)^3) \\
& - u_n - \Delta t(\lambda u_n + g(t_n)) \\
=\, & (u(t_n) - u_n) + \Delta t(\lambda u(t_n) + g(t_n)) - \Delta t(\lambda u_n + g(t_n)) \\
& + \frac{1}{2}(\Delta t)^2 u''(t_n) + \mathcal{O}((\Delta t)^3) \\
=\, & (1 + \lambda \Delta t)(u(t_n) - u_n) + \frac{1}{2}(\Delta t)^2 u''(t_n) + \mathcal{O}((\Delta t)^3).
\end{aligned}
\tag{10.18}
$$

Using the definition of the global error at t_n,

$$E_n = u(t_n) - u_n \tag{10.19}$$

and the quantity

$$\tau_n = \frac{1}{2}(\Delta t)u''(t_n) + \mathcal{O}((\Delta t)^2), \tag{10.20}$$

one obtains the recursive relation

$$E_{n+1} = (1 + \lambda \Delta t)E_n + \Delta t \cdot \tau_n. \tag{10.21}$$

Suppose we restrict our attention to the finite time interval $0 \le t \le T$, where $N = T/\Delta t$, and define

$$\|\tau\|_\infty = \max_{0 \le n \le N-1} |\tau_n|. \tag{10.22}$$

Then, from the recursive relation (10.21), one obtains[2] the bound for E_n,

$$|E_n| \le e^{|\lambda|T} T \|\tau\|_\infty = K\Delta t = \mathcal{O}(\Delta t) \text{ as } \Delta t \to 0, \tag{10.23}$$

for some positive constant K. Use has been made here of the fact that $\Delta t = T/N$. Equation (10.23) suggests that the FEM converges for solving (10.15) and is first-order accurate. Another way to state this is that, for a first-order accurate method, reducing the step size by a factor of two results in a decrease of the global error by roughly the same factor if Δt is small. It is always good to keep in mind, though, that this result is only valid asymptotically, as $\Delta t \to 0$.

---

[2] For detailed derivation, we encourage readers to see *Finite Difference Methods for Ordinary and Partial Differential Equations: Steady-State and Time-Dependent Problem* by Randall J. LeVeque, pp 139–140, SIAM Publish, 2007.

The code used to generate figure 10.1 is shown in its entirety in MATLAB script 10.1.

MATLAB Script 10.1: The FEM for a scalar linear initial-value problem

```
clear all;
close all;
N=5;
t=0:0.01:0.5;
exact = exp(-4*t)-t+2;
plot(t, exact,'-k')
hold on
for i=1:5
    t=0;
    u_n=3;
    dt=0.5/N;
    for k=1:N
        f_n=-4*u_n + 7-4*t;
        u_n=u_n+dt*f_n;
        t=t+dt;
        u(k)=u_n;
    end
    E_N(i)=exact(51)-u(N);
    up=[3,u];
    if (i==1)
    plot(0:dt:0.5,up, 'ks')
    elseif (i==2)
    plot(0:dt:0.5,up, 'kp')
    elseif (i==3)
    plot(0:dt:0.5,up, 'k+')
    elseif (i==4)
    plot(0:dt:0.5,up, 'kd')
    elseif (i==5)
    plot(0:dt:0.5,up, 'ko')
    end
    hold on
    N=N*2;
end
hold off
xlabel('t')
ylabel('u')
legend('exact solution', 'FEM solution, \Deltat=0.1', 'FEM
    solution, \Deltat=0.05', ...
        'FEM solution, \Deltat=0.025', 'FEM solution,
            \Deltat=0.0125', ...
        'FEM solution, \Deltat=0.00625')
axis([0 0.52 1.4 3.2])
E_N
```

10.5 Consistency, Stability and Convergence

A good look at figure 10.1 might raise an important question in the mind of a reader. While the numerical solution does approach the exact solution as the time step is reduced, in most real-life situations one doesn't know the exact solution. To have some degree of certainty that the numerical solution is valid one can clearly produce a sequence of approximations, with smaller and smaller time steps, until the difference between them becomes negligible. But many times the solution is needed sooner rather than later and this grid refinement study is not feasible. While there are various other mechanisms to tackle this problem, it should be clear that a related question may come to the reader's mind: since eventually the time step will have a non-zero, positive value, can that be too large for the numerical solution to diverge qualitatively in character from the exact solution? This turns out to be a difficult question to answer in general, so the following discussion will focus on the linear equations case, in particular equation (10.3). This is known as the *model equation* in the study of numerical methods for differential equations. For this model equation, the FEM produces the following sequence of approximations for a fixed time step Δt:

$$
\begin{aligned}
u_1 &= u_0 + \Delta t \cdot r u_0 = u_0(1 + r\Delta t), \\
u_2 &= u_1 + \Delta t \cdot r u_1 = u_0(1 + r\Delta t)^2, \\
&\vdots \\
u_n &= u_{n-1} + \Delta t \cdot r u_{n-1} = u_0(1 + r\Delta t)^n,
\end{aligned}
\tag{10.24}
$$

where $u_n \approx u(t_n) = u(t_0 + n\Delta t)$. Suppose this sequence is used to approximate the value of $u(T)$ at a final time $T = t_0 + N\Delta t$, where we can take $t_0 = 0$ without loss of generality. It is immediately obvious that one cannot take the limit $\Delta t \to 0$ without regard to other concerns. For example, if the number of time steps n is kept fixed, then $n\Delta t \to 0$ as well, so that $u(t_0 + n\Delta t) \to u_0$. Therefore, the number of time steps N used to obtain the approximation to $u(T)$ needs to be changed in direct connection with Δt. Consider then that $\Delta t = T/N$ and the numerical approximation is $u_N = u_0(1 + r\Delta t)^N$, which needs to be compared to the exact solution $u(T) = u_0 e^{rT}$. The final time T is here kept fixed, but $\Delta t \to 0$ and thus $N \to \infty$. Under these conditions, it is to be expected that $\lim_{N \to \infty} u_N = u(T)$. This is indeed the case:

$$
\begin{aligned}
\lim_{N \to \infty} u_N &= \lim_{N \to \infty} u_0 \left(1 + r\Delta t\right)^N \\
&= \lim_{N \to \infty} u_0 \left(1 + \tfrac{rT}{N}\right)^N = u_0 e^{rT},
\end{aligned}
$$

where use has been made of the fact that $\lim_{N \to \infty} (1 + \tfrac{a}{N})^N = e^a$. Thus, the FEM solution does converge to the exact solution at a fixed time T, at least for the model equation, in the limit $\Delta t \to 0$. Nevertheless, it turns out that the convergence of a numerical method for ordinary differential equations has two important ingredients: *consistency* and *stability*. They are both required in order for a numerical method that approximates the solution to a differential equation to converge.

Consistency is concerned with what happens with the approximation error over a single time step. If the error over a single time step doesn't go to zero as the size of the time step goes to zero, there is obviously little hope that the error that accumulates over a number of time steps will do so in the general case (although this may happen, for example by cancelation). Stated loosely, a numerical method for a differential equation is consistent if it faithfully reproduces the equation. To make this more rigorous, let us look at the FEM method and quantify the error introduced over a time step $\Delta t = h = t_{i+1} - t_i$, $i = 0, 1, 2, \ldots$. Under the usual assumptions over the smoothness of the solution $u(t)$, one can write:

$$u(t_{i+1}) = u(t_i) + \frac{du}{dt}\Big|_{t_i} (t_{i+1} - t_i) + \mathcal{O}(h^2),$$

which, using $u(t_i) = u_i$, can be rearranged upon division by h to show that

$$\frac{u(t_{i+1}) - u_i}{h} = f(u_i, t_i) + \mathcal{O}(h),$$

Consequently, since as we already know,

$$\frac{du}{dt}\Big|_{t_i} = \frac{u_{i+1} - u_i}{h} + \mathcal{O}(h),$$

and the sum of two terms of size $\mathcal{O}(h)$ is still of the same size, one obtains

$$\frac{u(t_{i+1}) - u_i}{h} = \frac{u_{i+1} - u_i}{h} + \mathcal{O}(h). \tag{10.25}$$

Multiplying through by a factor of h and doing the obvious simplifications leads then to the following relationship between the exact solution at time t_i and the numerical solution:

$$u(t_{i+1}) = u_{i+1} + \mathcal{O}(h^2).$$

Since we already know from equation (10.23) that the FEM is first-order accurate, we can see that this coincides with the difference between the numerical and the exact solution over a time step, $|u(t_{i+1}) - u_{i+1}|$, being $\mathcal{O}(h^2)$. This is very similar to what happens in the case of numerical quadrature, which is effectively what we do here; the error over one interval is one order of accuracy higher than the global error. It is convenient to define the *truncation error* of the FEM from equation (10.25) as the error in the approximation of the derivative, which is $\mathcal{O}(h)$. This obviously goes to zero as $h = \Delta t \to 0$, which means that the FEM is *consistent*.

For stability one can look at equation (10.23), which indicates that the global error of the FEM is bounded independently of h, as $h \to 0$. As a consequence, we say the FEM is stable asymptotically. Thus the FEM is both consistent and stable, and hence convergent, as $h \to 0$.

To further understand what stability means and why it is required, consider again the fact that a numerical method such as the FEM or the Runge-Kutta method will eventually be used with a finite-size, possibly constant time step. As the calculation proceeds, a random round-off error (which may easily arise due to finite precision) should not increase in such a way as to make the solution meaningless. Therefore, our numerical methods also need to be studied under the scenario that the time step is kept constant. Clearly, if the solution of the equation to be solved stays bounded for a finite time, the numerical solution should also remain bounded in that time frame. Addressing the stability problem in all depth is somewhat beyond the scope of this text. Nevertheless, a good understanding can be obtained by considering again the behavior of a method[3] for the solution of the model equation (10.3). To distinguish all possible situations, it is useful here to digress from our usual conventions and allow the rate of growth r to be a complex number, $r = \lambda + i\mu$. This means that our analysis will also cover equations of the form $du_R/dt = \cos(\mu t)$ and $du_I/dt = \sin(\mu t)$, which, when $\lambda = 0$, are the real- and imaginary-part, respectively, of the equation $du/dt = e^{rt}$ with $u(t) = u_R(t) + iu_I(t)$. In fact, if $r = \lambda + i\mu$, the solution of the latter differential equation has the form $u(t) = u_0 e^{\lambda t} [\cos(\mu t) + i \sin(\mu t)]$, so by allowing the constant r to be complex one can easily cover periodic solutions with the same model equation. Notice in this setting that if $\Re(r) = \lambda < 0$, then the amplitude of the exact solution decays with time. That is:

$$\left| \frac{u(t_2)}{u(t_1)} \right| = \left| \frac{e^{\lambda t_2} e^{i\mu t_2}}{e^{\lambda t_1} e^{i\mu t_1}} \right| = \left| \frac{e^{\lambda t_2}}{e^{\lambda t_1}} \right| < 1; \quad t_1 < t_2, \lambda < 0,$$

so that the exact solution necessarily remains bounded for all time under these conditions. This behavior should clearly be reproduced by any numerical method that is used to solve differential equations. It is therefore quite an unfortunate situation that this doesn't happen by default. In the case of the FEM, for example, the numerical approximation for two successive time steps produces:

$$\left| \frac{u_{n+1}}{u_n} \right| = \left| \frac{(1 + r\Delta t)^{n+1}}{(1 + r\Delta t)^n} \right| = |(1 + r\Delta t)|,$$

so in order to use the FEM method with confidence, the following condition needs to hold:

$$|1 + r\Delta t| < 1, \quad \text{for} \quad \Re(r) < 0.$$

---

[3]The type of stability discussed here is sometimes referred to as linear stability [8].

Let us start looking first at the case when $r \in \mathbb{R}, r < 0$, a particular case of the more encompassing $r = \lambda + i\mu$ in which $\mu = 0$. Then the condition becomes:

$$-1 < 1 + r\Delta t < 1, \quad r < 0.$$

The right-hand side of the equality poses no problems, as it reduces to $r\Delta t < 0$, which is automatically obeyed since the time step is always positive, $\Delta t > 0$, while $r < 0$ in the case under consideration. But the left-hand side leads to $-2 < r\Delta t$, which can be written as $\Delta t < (-2)/r$. Thus, the time step needs to be restricted; the larger the absolute value of the negative constant r is, the smaller the time step. This is known as a *stability condition*; for the FEM result to qualitatively reproduce the exact solution, such a condition must be obeyed.

In the more general case, with $r \in \mathbb{C}$, the stability condition for the forward Euler method is best written in the form:

$$|1 + z| < 1, \quad \text{for} \quad z = r\Delta t, \ \Re(z) < 0.$$

It helps our understanding if this is represented graphically, in the complex z-plane, with $\Re(z)$ on the horizontal axis and $\Im(z)$ on the vertical axis, as shown in figure 10.2. The ideal *stability region* for the model equation should be all the left half-plane (corresponding to $\Re(z) < 0$). Instead, the stability condition derived above is obeyed by the FEM only in a small disk with radius one and center at $z = -1$ inside the left-half plane. The case $r \in \mathbb{R}$ corresponds to the open interval that the disk cuts out of the horizontal axis, $-2 < z < 0$. One additional important remark can be made here: the linear stability region for FEM does not include points on the imaginary axis. That means that the method is not suitable for a purely oscillatory case, for example for an equation of the form $du/dt = \cos(t)$.

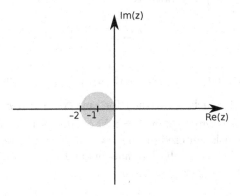

Figure 10.2: The linear stability region for the forward Euler method.

10.6   Explicit vs. Implicit Methods

Let us return to Euler's Method for the model equation $du/dt = ru$. Note that, when moving from t_n to t_{n+1}, we chose somewhat arbitrarily to evaluate the slope $f(u, t)$ at the point (u_n, t_n). Alternately, one can choose to evaluate it at (u_{n+1}, t_{n+1}) instead. This leads to the so-called *backward Euler method* (BEM):

$$u_{n+1} = u_n + \Delta t \cdot r u_{n+1} \tag{10.26}$$

for the model equation, or more generally:

$$u_{n+1} = u_n + \Delta t \cdot f(u_{n+1}, t_{n+1}) \tag{10.27}$$

As can be seen, the method obtained in this way is implicit. It makes sense to consider the possible disadvantage of using such a method, as well as its advantages.

▶ Disadvantage: looking at equation (10.27) above, one notices that it's not always easy to get u_{n+1} in terms of the known quantities t_n, u_n and t_{n+1}. If $f(u, t)$ is a nonlinear function, equation (10.27) defines u_{n+1} implicitly. A nonlinear equation must be solved to obtain explicitly the value of u_{n+1} for each time step. It is exactly for this reason that the BEM is known as an implicit method, while the FEM is an explicit method.

▶ Advantage: on the other hand, implicit methods do not usually have the stability restrictions on the time step that explicit methods tend to have. This can be rather easily seen by again looking at the same type of stability condition for the model equation. Expressing u_{n+1} from equation (10.26), one obtains:

$$\left| \frac{u_{n+1}}{u_n} \right| = \left| \frac{1}{1 - r\Delta t} \right|.$$

Asking now that, in the case of decay, $\left| \frac{u_{n+1}}{u_n} \right| \leq 1$, one gets the condition $\left| \frac{1}{1-r\Delta t} \right| < 1$ which is satisfied for any $\Delta t > 0$ if $r < 0$. The stability region in the complex plane $z = r\Delta t$ comprises this time the whole plane except the disk centered at the $z = 1$ and of unit radius. In particular, it completely covers the left half-plane $\Re r < 0$, which corresponds to decay in the analytical solution.

10.7 Multistep Methods

All the methods seen so far (FEM, BEM, Heun, RK4) are known as one-step methods. Even if they have several stages, as is the case for the latter two, they allow us to move from time t_n to time t_{n+1} without necessarily involving approximations for $u(t)$ at previous times. Another class of methods can be devised by using an interpolant for $f(u,t)$ that involves the values at time instances prior to t_n. For example, if t_{n-1} is also considered, one obtains a method of the form:

$$u_{n+1} = N(\Delta t, u_n, t_n, u_{n-1}, t_{n-1}),$$

where $N(\cdot)$ represents a numerical formula that necessarily involves evaluation of the right-hand side function $f(u,t)$.

Such methods are called *multistep methods*, as they require the knowledge of values of the dependent variable $u(t)$ one or more time steps prior to the current time t_n. These methods can be readily derived using our knowledge of interpolation. One can use the Lagrange form of the interpolant, or alternately Newton's divided differences, to approximate the integrand in equation (10.8). To clarify the process, let us look at an example of a *two-step method* that involves the values at t_{n-1} and t_n to advance the solution from t_n to $t_{n+1} = t_n + h$. The linear interpolant for the right-hand side function $f(u,t)$ can be written:

$$f(u(t),t) = f_{n-1}\frac{t-t_n}{t_{n-1}-t_n} + f_n\frac{t-t_{n-1}}{t_n-t_{n-1}} = f_{n-1}\frac{t-t_n}{-h} + f_n\frac{t-t_{n-1}}{h},$$

with an error $\mathcal{O}(h^2)$, where the obvious notation $f_j = f(u(t_j), t_j)$ for $j = n$ and $j = n - 1$ has been used. A straightforward calculation shows that

$$\int_{t_n}^{t_{n+1}} (t - t_n)dt = \frac{t_{n+1}^2 - t_n^2}{2} - ht_n = \frac{(t_n + h)^2 - t_n^2}{2} - ht_n = \frac{h^2}{2}$$

and similarly

$$\int_{t_n}^{t_{n+1}} (t - t_{n-1})dt = \frac{(t_n + h)^2 - t_n^2}{2} - h(t_n - h) = \frac{3h^2}{2}.$$

Therefore, the integral in equation (10.8) evaluates to:

$$\int_{t_n}^{t_{n+1}} f(u(t),t)dt = -\frac{f_{n-1}}{h}\frac{h^2}{2} + \frac{f_n}{h}\frac{3h^2}{2} + \mathcal{O}(h^3),$$

and consequently the following method, known as the *Adams-Bashforth two-step method*, is obtained:

$$u_{n+1} = u_n + \frac{h}{2}\left[3f(u_n, t_n) - f(u_{n-1}, t_{n-1})\right].$$

Since this method has a local error $\mathcal{O}(h^3)$, the global error incurred in advancing to a finite time is $\mathcal{O}(h^2)$. It is, therefore, similar to Heun's method in terms of accuracy. Unfortunately, the similarity ends here, as a consideration of what happens when taking the first time step clearly shows. To compute u_1, one would need both u_0 and u_{-1}, the latter of which is not available. Thus, the method is not self-starting; one needs to take a first step, from t_0 to t_1, using a single step method of the same accuracy (i.e. Heun). The computation can then proceed with the two-step method until the final time is reached.

Other methods in this category include the *Adams-Bashforth three-step method*, which has a local error $\mathcal{O}(h^4)$:

$$u_{n+1} = u_n + \frac{h}{12}\left[23f(u_n, t_n) - 16f(u_{n-1}, t_{n-1}) + 5f(u_{n-2}, t_{n-2})\right],$$

and the *Adams-Bashforth four-step method* with local error of order $\mathcal{O}(h^5)$,

$$
\begin{aligned}
u_{n+1} = \ & u_n + \tfrac{h}{24}\left[55f(u_n, t_n) - 59f(u_{n-1}, t_{n-1}) + 37f(u_{n-2}, t_{n-2})\right. \\
& \left. -9f(u_{n-3}, t_{n-3})\right].
\end{aligned}
$$

All these methods are not self-starting; one needs to use a one-step method of the same order of accuracy to compute enough steps before they can be used in stand-alone mode. For the model equation, they also tend to have a relatively small stability region and hence require more time steps to reach a given final time. On the other hand, the big advantage is that they are very accurate for the same amount of work when compared to one-step methods, since function values for $f(u, t)$ occuring on the right-hand side can be stored and subsequently reused over several time steps.

10.8 Higher-Order Initial Value Problems

Suppose now that the goal is to solve a second-order equation,

$$\frac{d^2u}{dt^2} = F\left(\frac{du}{dt}, u, t\right)$$

subject to a couple of initial conditions (ICs) of the form

$$
\begin{cases}
u(t = 0) = u_0; \\
\dfrac{du}{dt}(t = 0) = v_0;
\end{cases}
\quad (u_0, v_0 \text{ are provided values})
$$

This can be done easily if the second-order equation is recast as a system of first-order equations by introducing auxiliary variables as necessary. The following example shows what is involved.

Example Let us consider the initial value problem:

$$\begin{cases} \dfrac{d^2u}{dt^2} + 7\dfrac{du}{dt} + 3u = \cos(t), \\ u(0) = 1, \\ \dfrac{du}{dt}(0) = 0, \end{cases}$$

which describes a damped oscillatory system subjected to a periodic excitation. Then one can introduce the auxiliary variable $\frac{du}{dt} = v$; since $\frac{d^2u}{dt^2} = \frac{dv}{dt}$, one obtains following system of two equations that is equivalent with the initial second-order equation:

$$\begin{cases} \dfrac{du}{dt} = v \\ \dfrac{dv}{dt} + 7v + 3u = \cos(t) \end{cases}$$

or, upon rearrangement:

$$\begin{cases} \dfrac{du}{dt} = v \\ \dfrac{dv}{dt} = -7v - 3u + \cos(t). \end{cases}$$

These two coupled equations are now subject to the ICs $\begin{bmatrix} u(t=0) \\ v(t=0) \end{bmatrix} = \begin{bmatrix} 1 \\ 0 \end{bmatrix}$.

The system can be solved with the methods discussed so far, or by using MATLAB's own routines, which will be discussed below.

$$\|$$

10.9 Boundary Value Problems

Once one steps into the world of higher-order equations, some interesting questions appear. One of these is related to the fact that a second-order differential equation needs two ICs. It turns out, however, that instead of needing the two conditions (in the second-order case the value of the dependent variable and its derivative) imposed at the same point, some problems require imposition of two conditions at two different points. This type of problems are known as *boundary-value problems*; for most of them the independent variable is rather a space coordinate than time. For example, a common problem of this type occurs in the study of the deflection of a beam, where the two conditions at the endpoints may represent the fact that the beam is supported (known displacement) or clamped into a wall (displacement derivative known). Nevertheless, problems of this type for which the independent variable is still time arise in control theory, where a system needs to be adjusted to obtain a desired final state. An example of a second-order boundary value problem, written here still in terms of t as the independent variable, is:

$$\frac{d^2u}{dt^2} = F\left(\frac{du}{dt}, u, t\right); \quad u(t_a) = u_a, \ u(t_b) = u_b.$$

The methods developed for initial-value problems can be readily extended to solve boundary-value problems of this type through the use of the *shooting* technique. The idea is to recast the problem as an initial-value problem of the form

$$\frac{d^2u}{dt^2} = F\left(\frac{du}{dt}, u, t\right); \quad u(t_a) = u_0, \ u'(t_a) = v_a, \tag{10.28}$$

where the initial slope v_a is not known. The problem can then be stated as follows: find the value of the initial slope $v_a = v_a^*$ such that the solution of the initial-value problem

$$\frac{d^2u}{dt^2} = F\left(\frac{du}{dt}, u, t\right); \quad u(t_a) = u_a, \ u'(t_a) = v_a^*$$

satisfies $u(t_b) = u_b$. Since, upon integrating equation (refeq:shoot1), the solution at the final time can be seen to be only a function of the assumed initial slope, a fact that can be denoted by $u(t_b) = U(v_a)$, this can be seen to lead to a nonlinear equation of the form $T(v_a) = U(v_a) - u_b = 0$. The solution of this equation, v_a^*, has the property that $T(v_a^*) = 0$, hence $U(v_a^*) = u_b$. It is helpful to think about $T(v_a)$ as a *target function*. Hitting the target entails adjusting the initial slope, i.e. pointing a fictitious gun in the (t, u) plane higher or lower, until the good orientation is found. To find the zero of the target function one can obviously make use of the methods for solving nonlinear equations, of which the bisection method is straightforward to apply, as it doesn't require the derivative of the target function.

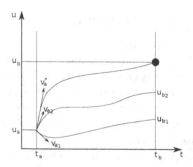

Figure 10.3: Schematic of the shooting method.

10.10 MATLAB® Built-in Functions

MATLAB offers several functions that can be used to solve differential equations or systems thereof. Only two of these are discussed here. The first one, ode23, provides a second-order Runge-Kutta method that is embedded within with a third-order method. The embedding offers a way to approximate the error and adjust the time step accordingly. The second method, ode45, is similar but uses a fourth-order embedded method. Both have the same conventions for their arguments as highlighted in the following examples.

   Example To solve the single differential equation:

$$\begin{cases} \frac{du}{dt} = \sin(t) \\ u(0) = -1 \end{cases} \quad \text{for } t \in [0, 10]$$

using the second-order method, one can define the right-hand side as an anonymous function of two variables (in the code below the $0 \cdot y$ is just included for clarity). This function is then passed as an argument to ode23, together with the time span and the value of the initial condition. The time span is set up as a row vector containing the initial and final times, as shown below. The output consists of two vectors, one containing a time discretization of the time span and the other the corresponding values for the dependent variable at these time moments.

MATLAB Script 10.2: Using ode23

```
f = @(t,y) sin(t)+0*y;           % RHS
[tout,yout] = ode23(f, [0,10], -1);    % solve ode
plot(tout, yout)                  % plot solution
```

$$\parallel$$

   Example To solve the system

$$\begin{cases} \dfrac{dy_1}{dt} = y_2 \\ \dfrac{dy_2}{dt} = -7y_2 - 3y_1 + \cos(t) \end{cases} \quad \text{with initial conditions} \quad \begin{bmatrix} y_1(0) \\ y_2(0) \end{bmatrix} = \begin{bmatrix} 1 \\ 0 \end{bmatrix}$$

one first needs to write a function file, say myode2.m, containing the definition of the right-hand side terms in the two equations. These must be arranged in a function column vector:

Matlab Function 10.1: Defining the RHS for a system of ODEs

```
function [yprime]=myode2(t,y)

yprime=[y(2); -7*y(2)-3*y(1)+cos(t)]

end
```

To use this definition, one needs to pass the function to the MATLAB built-in solver. The example below uses the **ode45** method to solve the system, then plots the first component, y_1, of the solution.

MATLAB Script 10.3: Using ode45 for a 2x2 system

```
[t,y]=ode45('myode2',[0 1],[1;0]);    % solve 'myode2'
plot(t, y(i,1));                       % plot y(1) vs. t
```

$$\|$$

10.11 Exercises

Ex. 1 For each one of the differential equations below, check which ones of the functions listed following them are solutions on the provided intervals.

(a)

$$\frac{du}{dt} = 5t - \frac{3}{t}u, \quad t \in (0, \infty); \quad u_1(t) = t^3 - 2; \quad u_2(t) = t^2$$

(b)

$$\frac{du}{dt} = \cos(t), \quad t \in (-\infty, \infty); \quad u_1(t) = \sin(t); \quad u_2(t) = 2 + \sin(t)$$

(c)

$$\frac{du}{dt} = \frac{2t^3}{u+2}, \quad t \in (0, \infty); \quad u_1(t) = t^2 - 2; \quad u_2(t) = t^3$$

Ex. 2 Solve numerically the logistic differential equation for population dynamics, equation 10.4, for the values $p_0 = 1$, $r = 0.4$, $K = 6$. Write a code that implements the forward Euler method and use this code with a time step $\Delta t = 0.1$ to advance the solution to $t = 5$. Prove that the analytical solution given in Ex. 14 of section 3.11 is correct, then compare your numerical solution with this analytical solution for the provided initial condition and parameters.

Ex. 3 Rewrite the second-order equation

$$y'' - t^2 y' + ty = t^2 - 1; \quad y(0) = 1, y'(0) = 0$$

as a system of two first-order equations. Solve this system for $t \in [0, 2]$ using the forward Euler method with a time step $\Delta t = 0.1$.

Ex. 4 Solve the system of equations in the above problem using the MATLAB function `ode45`.

Ex. 5 Write a MATLAB function that implements the classical fourth-order Runge-Kutta method to solve an initial-value problem involving a single differential equation. Use this together with the forward Euler method to solve the differential equation:

$$\frac{du}{dt} = \cos(t), \quad u(0) = 0, \quad t \in [0, 15]$$

Use a time step $\Delta t = 0.1$. Plot the resulting numerical values obtained from the two methods versus the exact solution $u(t) = \sin(t)$ on the same plot. Use markers (for example stars or circles) for the numerical data points. Do you notice any difference between the two solutions? Can you explain what happens in the case of the forward Euler method?

Ex. 6 Write a function that implements the classical Runge-Kutta fourth-order method for the case of a system of four equations with four unknowns. Use this function to solve the following linear system of differential equations, which describes a set of elastic springs and masses coupled in series:

$$x' = u; \quad y' = v; \quad u' = -2x + (3/2)y; \quad v' = (4/3)x - 3y$$

subject to the initial conditions $x(0) = -1$, $y(0) = 4$, $u(0) = 1$, $v(0) = 1$. Choose a step size $\Delta t = 0.01$ and use your solutions to:

(a) Plot the values of the four variables as a function of time, each on a different plot, up to $t = 20$. These are the time series plots.

(b) Plot $u(t)$ (vertical axis) versus $x(t)$ (horizontal axis) for the same range in t. This is an example of a *phase plot*, which is very useful for obtaining information about dynamical systems. In particular, if the phase plane solution is a closed curve, the system exhibits periodic oscillation.

Ex. 7 The nonlinear van der Pol equation

$$\frac{d^2 x}{dt^2} + \mu(x^2 - 1)\frac{dx}{dt} + x = 0,$$

describes the evolution of the current $x(t)$ in an electric triode, a type of vacuum tube that was used to amplify the received signal in older electronic devices. For large values of x and $\mu > 0$, the system is damped, since the dx/dt term has a positive coefficient; the converse is true for small values of x. The resulting $x(t)$ exhibits therefore a *limit cycle*: an oscillation that either increases or decreases in amplitude to eventually become periodic. Solve the equation by rewriting it as a first-order system. With $v(t) = dx/dt$, use the value $\mu = 0.25$ and two different sets of initial conditions: (a) $x(0) = 0.1$, $v(0) = 0.1$; (b) $x(0) = 4$, $v(0) = -2$. Advance the two solutions to $T = 100$ using the RK4 method and $\Delta t = 0.05$. Plot $x(t)$ versus t for both solutions, then do again a phase plot: plot $v(t)$ versus $x(t)$ for both data sets in the same figure. Notice how the phase plot clearly shows the limit cycle as the closed curve on which both data sets eventually settle.

Ex. 8 The solution of the nonlinear Lorenz system of equations, which exhibits chaotic behavior, has been studied extensively. This three-by-three system is given by:

$$\begin{cases} \dfrac{dx}{dt} = \sigma(y - x) \\[2mm] \dfrac{dy}{dt} = x(\rho - z) - y \\[2mm] \dfrac{dz}{dt} = xy - \beta z \end{cases}$$

where the constants σ, ρ and β are all positive and govern the behavior of the system. Your goal in this exercise is to study the behavior of the solution by integrating the system in time; use $\sigma = 10$, $\rho = 28$ and $\beta = 8/3$ for the parameters and the RK4 method to integrate in time up to $T = 100$. Use the initial conditions $x(0) = 0.1$, $y(0) = 0.1$ and $z(0) = 0.1$ and time steps $\Delta t_1 = 0.1$ and $\Delta t_2 = 0.05$. Create the three phase plots $x(t)$ versus $y(t)$, $x(t)$ versus $z(t)$ and $y(t)$ versus $z(t)$ for both cases.

Ex. 9 Solve the system of predator-prey equations

$$\begin{cases} \dfrac{dx}{dt} = 0.9x - 0.45xy \\[2mm] \dfrac{dy}{dt} = -0.8y + 0.3xy \end{cases}$$

subject to the initial conditions $x(0) = 2$ and $y(0) = 1$.

(a) Solve using Euler's forward method and a step size $\Delta t = 0.1$. Advance the solution, if possible, up to $T = 50$. Plot the resulting $x(t)$ and $y(t)$ versus t; also create the phase plot $y(t)$ versus $x(t)$. Repeat for $\Delta t = 0.01$.

(b) Solve using the Runge-Kutta order 4 method (RK4), again with $\Delta t = 0.1$ and $\Delta t = 0.01$. Plot again the functions as above. Explain why you think there are noticeable differences in the results obtained by these two methods.

Ex. 10 Consider the scalar initial-value problem

$$y' = \kappa(y - \cos t) - \sin t,$$
$$y(t_0) = y_0. \tag{10.29}$$

(a) Verify by substitution that

$$y(t) = e^{\kappa(t-t_0)}(y_0 - \cos(t_0)) + \cos t$$

is the solution to the initial-value problem (10.29).

(b) Let $\kappa = -20$ and the initial condition $y(0) = 3/2$. Apply the FEM with the time step $\Delta t = 0.1$ for solving (10.29) up to a final time $T = 3$. Plot the computed solution against the exact solution for $0 \le t \le 3$.

(c) Repeat part (b), but using the time step $\Delta t = 0.05$. Compare your computed solution with that of part (b).

(d) The implicit, backward Euler method (BEM) for a linear differential equation

$$\frac{dy}{dt} = \kappa y + g(t)$$

is given by

$$y_{i+1} = y_i + \Delta t \cdot (\kappa y_{i+1} + g(t_{i+1})), \quad i = 0, 1, 2, 3, \ldots, N - 1,$$

where $N = T/\Delta t$ and T is a prescribed final time. The above equation is equivalent to

$$(1 - \kappa \Delta t)y_{i+1} - y_i = \Delta t g(t_{i+1}), \tag{10.30}$$

Given an initial condition $y(t_0) = y_0$, (10.30) forms an $N \times N$ lower-triangular linear system for the unknown vector $\vec{y} = [y_1, y_2, \ldots, y_N]^T$. Apply the BEM to (10.29) with the time step $\Delta t = 0.2$. Compare your computed solution with that of part (b) and (c). For $\kappa < 0$, if $-\kappa$ is large, (10.29) is an example of a so-called *stiff problem*[4].

---

[4]A stiff problem exhibits a rapid transition in some period of its development. In our case, the initial condition produces a sharp gradient in the solution curve for large $-\kappa$. Hence the example is a stiff problem. An explicit method, such as the FEM, requires a small time step to capture such rapid transition. The BEM is the first member of the backward differentiation formula (BDF) family that is a class of implicit methods suitable for solving stiff problems. A BDF method can use a far larger time step than an explicit method without losing stability.

Ex. 11 The usual BEM formulation for the general first-order differential equation

$$\frac{du}{dt} = f(u, t)$$

is given by

$$u_{n+1} = u_n + (t_{n+1} - t_n) \cdot f(u_{n+1}, t_{n+1}).$$

Use this method to solve numerically the differential equation

$$\frac{du}{dt} = \frac{4t - t^3}{4 + u^3}$$

subject to the initial condition $u(-3) = 0.4962108225$. Write a code that implements the method with a constant time step $\Delta t = t_{n+1} - t_n = 0.05$ and advance the solution to $t = 3.25$. Repeat the same calculation with the RK4 method and compare the results. Note that for BEM you will need to solve a nonlinear equation at each time step.

Ex. 12 Solve the boundary-value problem

$$u'' = \frac{3}{326} \left(uu' - 66u - 90u' - 3t^5 + 25t^4 \right) - \frac{2180}{326}$$

with $u(0) = 0$ and $u(3) = -6$ for $0 \le t \le 3$, using the shooting method with RK4 as the main integration engine.

Appendix A

Calculus Refresher

This appendix states several results from Calculus that are used to derive numerical methods. Proofs are not provided since those interested in them might want to consult an in-length exposition in a devoted Calculus textbook [18, 13]. The focus here is on how these results are used in the numerical framework.

A.1 Taylor Series

Definitely some of the most important and often used results in calculus, from a numerical methods perspective, are those pertaining to Taylor series. These results involve the approximation of functions by easy-to-compute polynomial sequences and provide us, in turn, with a way to investigate the errors involved in numerical approximations. The following theorem is widely used throughout this text:

Theorem A.1.1 (Taylor series with remainder): Suppose the function $f(x)$ has $n + 1$ continuous derivatives on an interval $[a, b]$ and choose a fixed point $x_0 \in [a, b]$. Then for all $x \in [a, b]$, the value of $f(x)$ can be expressed as

$$f(x) = \sum_{k=0}^{n} \frac{(x - x_0)^k}{k!} f^{(k)}(x_0) + R_{n+1}(x),$$

where

$$R_{n+1}(x) = \frac{(x - x_0)^{n+1}}{(n + 1)!} f^{(n+1)}(\xi)$$

for some point $\xi \in [a, b]$ that depends on x, $\xi = \xi(x)$, located between x and x_0.

The expression for $f(x)$ in this theorem is referred to as the Taylor series for $f(x)$ centered at x_0. When the point $x_0 = 0$, the series thus obtained for $f(x)$ is known as the *Maclaurin series*.

In the derivation of numerical methods, the Taylor series is written in several different ways, depending on the context. A very common way is to

DOI: 10.1201/9780429262876-A

let $x = x_0 + h$ and therefore obtain

$$\begin{aligned}
f(x) &= \sum_{k=0}^{n} \frac{h^k}{k!} f^{(k)}(x_0) + R_{n+1}(x) \\
&= f(x_0) + h f'(x_0) + \frac{h^2}{2} f''(x_0) + \frac{h^3}{6} f'''(x_0) + \dots
\end{aligned}$$

where the difference $x - x_0 = h$ is thought to go to zero in the limit. There is a strong connection between this form and the ubiquitous use of the big-oh notation in numerical analysis. Indeed, notice that by using the successive powers of h as gauge functions, one can write:

$$\begin{aligned}
f(x) &= f(x_0) + O(h) \\
&= f(x_0) + h f'(x_0) + O(h^2) \\
&= f(x_0) + h f'(x_0) + \frac{h^2}{2} f''(x_0) + O(h^3)
\end{aligned}$$

and so on. It may also help to notice here the basic rules for working with the big-oh notation: $O(ch^n) = cO(h^n) = O(h^n)$ for any integer n and constant c and $O(h^n) + O(h^m) = O(h^n)$ for any integers n, m such that $n \leq m$.

A.2   Riemann Integrals

The numerical quadrature techniques that are discussed in Chapter 9 make implicit use of the definition and the proof for the existence of the Riemann integral, summarized here. Define first a partition P of the interval $[a, b]$ as a set of points $P = \{x_0, x_1, \dots, x_n\}$ such that $a = x_0 \leq x_1 \leq x_2 \dots \leq x_n = b$. For a given partition P, define also its norm as being $||P|| = \max_{i=1,2,\dots,n} \Delta x_i$, with $\Delta x_i = x_i - x_{i-1}$. Then the Riemann integral of the function $f(x)$ on the interval $[a, b]$ is defined by

$$\int_a^b f(x)dx = \lim_{||P|| \to 0} \sum_{i=1}^{n} f(\bar{x}_i)\Delta x_i,$$

provided that the limit exists and is the same for all partition choices. Here the points $\bar{x}_i$ are points such that $\bar{x}_i \in [x_{i-1}, x_i]$. The following theorem addresses the existence of the Riemann integral and provides a means for its evaluation.

Theorem A.2.1 (Riemann Integral): Suppose the function $f(x)$ is continuous on the interval $[a, b]$. Then its Riemann integral on $[a, b]$ exists and is given by:

$$\int_a^b f(x)dx = \lim_{n \to \infty} \frac{b-a}{n} \sum_{i=1}^{n} f\left(a + \frac{b-a}{n}i\right).$$

A.3   Other Important Results

***Theorem* A.3.1** (Rolle's Theorem)*:* Suppose the function $f(x)$ is continuous on the interval $[a, b]$, differentiable on (a, b) and also that $f(a) = f(b)$. Then there exists at least one point $\xi \in (a, b)$ such that $f'(\xi) = 0$.

***Theorem* A.3.2** (Main Value Theorem)*:* Suppose the function $f(x)$ is continuous on the interval $[a, b]$ and is differentiable on (a, b). Then there exists at least one point $\xi \in (a, b)$ such that

$$f'(\xi) = \frac{f(b) - f(a)}{b - a}.$$

***Theorem* A.3.3** (Extreme Value Theorem)*:* Suppose the function $f(x)$ is continuous on the interval $[a, b]$. Then there exist two points $x_m, x_M \in [a, b]$ such that $f(x_m) \leq f(x) \leq f(x_M)$ for all $x \in [a, b]$. If $f(x)$ is also differentiable on (a, b), then the points x_m, x_M occur either at the endpoints of the interval $[a, b]$ or where the derivative of $f(x)$ is zero. The point x_m is referred to as the *absolute minimum* and the point x_M as the *absolute maximum* of the function $f(x)$ on $[a, b]$.

***Theorem* A.3.4** (Intermediate Value Theorem)*:* Suppose the function $f(x)$ is continuous on the interval $[a, b]$ and let m be any value between $f(a)$ and $f(b)$. Then there exists at least one point $\xi \in (a, b)$ such that $f(\xi) = m$.

Appendix B

Introduction to Octave

B.1 The Problem of Choice

The case for Octave is very simple: its interface is similar to MATLAB, while the syntax is almost identical and getting ever closer with each new release. The big advantage, however, is that it is free software. Octave has been developed under the open-source paradigm and a large number of contributors have come from the international mathematics community. The advantage that MATLAB still offers over Octave is a more active support, faster update cycles and a larger number of features. For example, as of this writing, the current Octave version does not have an implementation of the `arcsin` function, whereas MATLAB does. As another example, plotting from MATLAB may be easier while offering a larger number of features and export options. Nevertheless, for most computational applications, the features that Octave still misses will most probably not create any problems. Octave is probably at its best when used on a Linux computer, its native platform, although there are only minor differences with respect to the Windows and MacOS interfaces.

B.2 Octave Basics

To work with Octave, one can largely follow the indications for MATLAB code in this book. Octave offers both a command-line interface that is very fast and is similar to the REPL scripting capabilities of Julia or Python and a graphical interface that is similar but slightly different from MATLAB. The latter doesn't offer all the options, but the basic functionality (the text editor and the command window) is available. A snapshot of the graphical interface is shown in figure B.1.

DOI: 10.1201/9780429262876-B

Figure B.1: The Octave graphical interface (GUI).

B.3   Octave Code Examples

There are a few things to be aware of when writing code directly in Octave. The immediate difference that one becomes aware of when using the Octave editor is in the form of the termination commands that replicate the **end** keyword in MATLAB. Octave uses specific commands, depending on the type of **end** involved. For example, to end a **for** loop, the Octave editor will, by default, insert an **endfor** instead of **end**. However, if the **endfor** is replaced by a simple **end**, the code is still completely functional. Thus, importing files that have been edited and run in MATLAB does not raise any problem, while files that have been created in Octave might need some minor editing to run in MATLAB.

The following function computes the Lagrange interpolant $L_k(x)$ using the data points xdata. It corresponds to the MATLAB code shown in 6.3. Notice the particular **end** keywords for ending the loops and the function, respectively.

Octave Script B.1: Lagrange Interpolant

```
function [val] = lagrange( k, x, xdata)
  val = 1.0;
  N = length(xdata);
  for j = 1:(k-1)
    val = val * ( x - xdata(j) ) / ( xdata(k) - xdata(j) );
  endfor
  for j = (k+1):N
    val = val * ( x - xdata(j) ) / ( xdata(k) - xdata(j) );
  endfor
endfunction
```

Appendix C

Introduction to Python

C.1 The Problem of Choice

Python is a language that has surged in popularity in the last few years and has widely replaced Java as the first language taught to college computer science majors. The interest it has generated is due to a number of factors, among which the foremost is probably its readability: the language was designed with this as a major goal. Other advantages include its scripting capabilities, the dynamic typing model, its easy interaction with other main languages and the large number of available libraries. Those who learn Python will have the advantage of mastering a very general-purpose language that is still widely used in the computational science community where it has become the lingua franca for machine learning and data science.

Notwithstanding its ease of use, to obtain some of the capabilities that MATLAB offers Python enthusiasts must make informed use of several libraries that need to be imported manually. While low-level commands or functions in these libraries have names that are close or totally identical to those in MATLAB, there are many differences. On the other hand, most of the higher-level functions are not available immediately and need to be written by the user (that is, in fact, one reason why learning computational methods is still important). Python is also developed under an open-source resource model, which means that documentation and assistance may not be available immediately. Nevertheless, many free tutorials are available over the Internet, and help is usually just a Google search away. The enthusiasts interested in using Python should consult the very readable introduction to the language by Downey [4] as well as the coverage of the use of Python for numerical techniques by Kiusalaas [11].

C.2 Python Basics

Python differs from MATLAB in a number of ways. While not all the differences can be mentioned in a short appendix, the aim here is to give enough

DOI: 10.1201/9780429262876-C

information so that someone can start using Python right away if previous knowledge of MATLAB is assumed. Python can be deployed in a variety of ways. It can be run as a stand-alone application in a terminal window, where commands are entered interactively (also known as a REPL, an abbreviation for read-evaluate-print-loop or CLI, for command-line interface) as shown in figure C.2. Such a REPL environment can actually be obtained in a web browser without any local installation of Python, by just visiting the website `www.repl.it`. Nevertheless, probably the most widely used way to use Python is by writing scripts, which are stored as plain text files and sent to the Python engine to be interpreted and executed. Somewhat more recently, the use of Python in a notebook interface has also become widespread; in such an environment, units (usually called cells) of Python script are executed in succession and can be interspersed with cells that contain plain or mathematical text. Notebooks have been used as interfaces to computer-algebra systems such as Mathematica and Maple for some time, as they offer a great way to convey ideas. The most widely used notebook interface to Python is currently Jupyter, pictured in figure C.1.

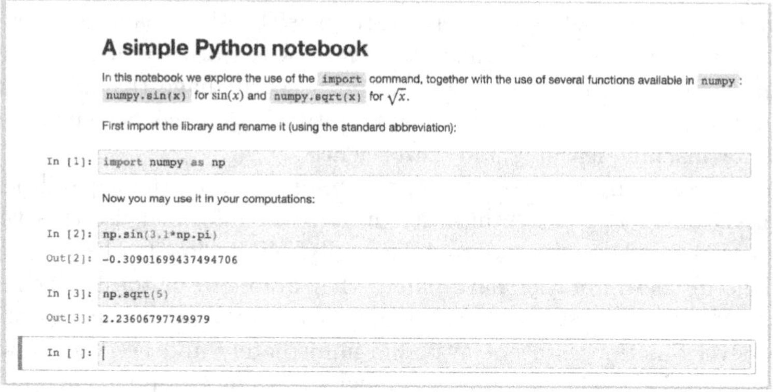

A simple Python notebook

In this notebook we explore the use of the `import` command, together with the use of several functions available in `numpy` : `numpy.sin(x)` for $\sin(x)$ and `numpy.sqrt(x)` for $\sqrt{x}$.

First import the library and rename it (using the standard abbreviation):

```
In [1]: import numpy as np
```

Now you may use it in your computations:

```
In [2]: np.sin(3.1*np.pi)
Out[2]: -0.30901699437494706

In [3]: np.sqrt(5)
Out[3]: 2.23606797749979

In [ ]:
```

Figure C.1: The Jupyter interface.

For historical reasons, two important versions of Python are still in use today: Python 2 and Python 3. The latter was introduced as a major overhaul of the former and is therefore not backwards compatible. There still exist many Python 2 scripts, distributed over the Internet or otherwise, that perform useful tasks and have not been ported to Python 3 yet. Nevertheless, Python 2 will not be supported any more as of January 2020. Python 3 will be assumed for the rest of the presentation here.

Most computer systems do have a version of Python already installed and used for various tasks by the operating system. To avoid interference with this version, the use of a packaged Python distribution is recommended, the most popular one being probably the Anaconda distribution. This will install a virtual environment that shields the base Python installation on your machine from interaction with later versions. Alternately, one can create their own

Figure C.2: Python REPL.   **Figure C.3:** The Spyder IDE.

virtual environments, with a different Python version in each. This avoids potential conflicts between pieces of software that use different versions.

Once a Python version is installed, most users will prefer to also install an integrated development environment (IDE) that allows you to edit and debug Python code. Some of these IDEs will in fact allow you to install Python as well, in which case it is not necessary to do that beforehand. Two of the free popular IDEs are Spyder (part of the Anaconda distribution, pictured in figure C.3) and the community/education editions of PyCharm, developed by JetBrains. For academic users, Enthought Canopy offers another free, very well integrated alternative.

C.3   Installing Python

Installing and using a Python work environment will depend on the chosen platform and is subject to change in time. Therefore, only a brief mention of the installation for the Anaconda distribution is made here. Currently the installers for all main computing platforms can be downloaded from the `anaconda.org` website. The installer sets up Python, together with the Spyder IDE, the Jupyter notebook interface and a number of other tools widely used for computational and data science, such as visualization software. it also creates a base environment to use these tools without interfering with the system-wide Python engine.

C.4   Python Code Examples

C.4.1   Libraries

In order to perform most scientific computations, the use of Python's numerical library NumPy is necessary. This library defines, among others, the functions needed when working on real numbers as well as the usual constants. It has to be imported into the present session for these definitions to become available. After the library is imported, the variables and functions defined within can be accessed by prepending the name of the function with the name of the library followed by a dot:

```
import numpy
numpy.sin(3.21*numpy.pi)
```

Other libraries that are very useful for scientific computation are SciPy and MatPlotLib. Because writing the name of the library every time a function is used can easily become inconvenient, Python programmers have developed a rather standard set of abbreviated names for the libraries, which can be used instead of the actual name as shown below:

```
import numpy as np
np.sqr(3)
```

C.4.2   Native Python arrays: lists and tuples

It is when using arrays that many users find Python concepts somewhat different from other languages. Python doesn't offer off-hand multi-dimensional arrays; instead, the users have to craft their own arrays out of the basic one-dimensional array type. This is a very general list of elements of any type, with index numbering starting at zero (a stark contrast with MATLAB, where indexing starts at one). To create a Python list just enclose the elements, separated by commas, in between square brackets. A tuple will likewise be created if parentheses are used instead.

```
L = [ 1.0, 1.1, 1.2, 'abc' ]
T = ( 1.1, 1.2, 1.3, 1.4)
print( L[1] )
```

The main difference between a list and a tuple is that the first is *mutable*, meaning that its elements can be modified after its declaration, while tuples are *immutable*.

Several functions, similar to those in MATLAB, can be used on lists and tuples that have only numerical entries. These include `max`, `min`, `sum` and so on. Multi-dimensional arrays can be created, for example, as lists of lists; they can be ragged, in the sense that the various lists they are made up from do not need to have the same length:

```
L = [   [1.0, 1.1, 1.2], [3, 4, 5], ['a', 'b']   ]
T = (   (1.1, 1.2, 1.3), (1.4, 1.5) )
print( L[1] )   # output: [3, 4, 5]
print( L[2] )   # output : ['a', 'b']
```

Using this kind of arrays would require creating our own functions for their manipulation. The NumPy library provides instead specialized numerical versions of these arrays for which the familiar vector and matrix operations are defined.

C.4.3 NumPy arrays

Upon importing NumPy, the user has access to arrays that can only store numerical data type but for which the usual operations are already defined. The syntax is inherited from the Python list syntax:

```
r = np.array( [   1.0, 2.0, 3.0   ] )   # row vector
c = np.array( [ [3], [4], [5] ] )   # column vector
k = np.transpose( r )   # column vector
M = np.array( [ [1, 2], [3, 4] ] )   # 2x2 matrix
print( M[0,0] )   # output : 1
print( M[0] )   # output: [1, 2]
```

Very similar to MATLAB, Python offers a way to extract only a subset of elements from an array with the *slicing* syntax r[start:end:step]; notice however that the order of the arguments is different. There are a number of functions that create the most usual arrays; zeros((m,n)), ones((m,n)) and linspace(a,b,n) all do what you expect them to do. For any previously constructed NumPy array A, we can also obtain information on its shape and size using the predefined functions A.shape (which returns the dimensions) and A.size (which returns the total number of elements).

In contrast to MATLAB, the library doesn't offer the convenient infix notation for array operators. For example, the dot product of two vectors in Numpy is performed with the function numpy.dot and matrix multiplication with numpy.matmul. Nevertheless, the usual operations on vectors and matrices available through NumPy are highly optimized (they provide only an interface to numerical linear algebra functions written in very performant, compiled languages such as C). The code excerpt below, a Python version of the MATLAB code 6.1, is an example of their use in computing the coefficients of a straight-line fit and should be self-explanatory.

```
import numpy as np

x = np.array([0, 2.6, 4.95, 7.02])
y = np.array([0, 5, 10, 15])
sumx = sum(x)
xy = np.multiply(x,y) #element-wise product
x2 = np.square(x)
M = np.array([[sum(x2), sumx], [sumx, len(x)]])
rhs = np.array([[sum(xy)], [sum(y)]])
sol = np.linalg.solve(M,rhs)
print(sol)
```

C.4.4   Loops and conditionals

One of the main distinguishing features of Python is the lack of delimiters for its control structures; the ubiquitous **end** appearing in MATLAB code is gone. To compensate for this, the language uses a strict indentation policy in order to delineate blocks of code. Throughout this book the for loop was widely used for repetitive tasks. In Python, the most usual form of this control structure is

```
for k in range(start, end, step) :
   # repeated things
   # to do
# This is not in the loop any longer
```

Notice the presence of the colon at the end of the line declaring the loop; also notice the **range** function, which one can think of as creating a vector containing the iteration counters. Just like the colon operator in MATLAB, not all arguments are needed for this function. The indented block of code is executed for each value in this vector. It is important to always remember that Python starts array indexing at zero:

```
A = np.zeros((5))
for k in range(5):
    A[k] = k
print(A)   # output [0. 1. 2. 3. 4.]
```

Until someone with a mathematical background gets used to this peculiarity, **while** loop may be more desirable, since it makes the condition governing the loop easier to read for a mathematician's eye:

```
B = np.zeros((5))
k = 0
while k < 5:
    B[k] = k
    k = k + 1
print(B)  # output [0. 1. 2. 3. 4.]
```

The **if** conditional has a similar form as the loops, again using the colon and indentation to delineate its scope:

```
for j in range(1:7):
  if (j == 21):
    print('This should not happen.')
  elif j % 2 == 0: # check if divisible by 2
    print(j, ' is an even number.')
  else:
    print(j, ' is an odd number.')
```

The logical and relational operators in Python, listed in table C.1, are again very similar to MATLAB, but not totally identical. As a last remark in this section, notice the use of the **print** function instead of MATLAB's **disp**. For the most part **print** is easy to use, although it does take a variety of forms that will not be covered in all detail here.

TABLE C.1: Python Relational and Logical Operators.

Infix	Operator
==	equal to
!=	not equal
>	greater than
<	less than
>=	greater than or equal to
<=	less than or equal to
and	logical AND
or	logical OR

C.4.5   Python functions

Similar to MATLAB, Python does offer a way to define very simple, one-line, functions that are known as **lambda** functions, as well as more complicated functions that need a longer block of code. The Python syntax again makes use of the colon to introduce the code for the evaluation of the function's value. Both kinds are exemplified in the code excerpt below.

```
f1 = lambda t: t**2 + 1
def f2 (x):
   return x**2 + 1

print(f1(2)) # output: 5
print(f2(3)) # output: 10
```

C.4.6   Data visualization

A widely used Python library is MatPlotLib, which contains the useful `PyPlot` package. The code below re-creates the same plot shown in figure 3.2, this time in a Python environment.

```
import numpy as np
import matplotlib.pyplot as plt

day = np.array([1, 2, 3, 4, 5, 6, 7])
temp = np.array([71, 72, 63, 64, 64, 70, 73])

plt.figure()  # create a figure handle
plt.plot(day,temp) # plot data
plt.xlabel('Day')
plt.ylabel('Temperature')
plt.title('Basic Plot Example')
```

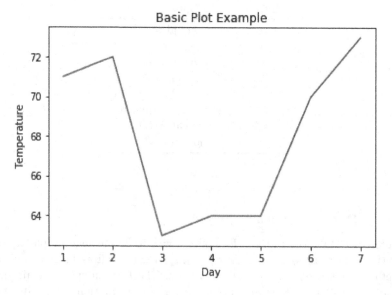

Figure C.4: Basic plot with MatPlotLib.

To create a different plotting frame one needs to call the `figure` function again, while in order to plot several data sets in the same figure one needs to use additional calls to `plot` with the appropriate data. In the latter case, a legend is usually needed and can be created easily:

```
import numpy as np
import matplotlib.pyplot as plt

x = np.linspace(0,10,5) # data points
plt.figure() # create figure
plt.plot(x, x,linewidth=2) # call plot functions
plt.plot(x, x+2, linewidth=2)
plt.plot(x, x+4, linewidth=2)
plt.title("Default MatPlotLib colors", size = 14, weight = '
                           bold')
plt.legend(('1st','2nd','3rd'))
```

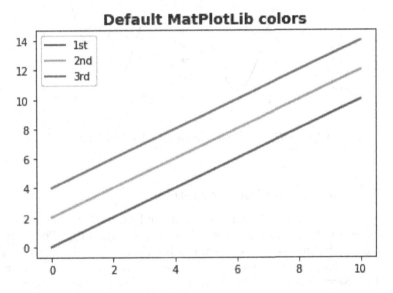

Figure C.5: Several data sets plotted with MatPlotLib.

Many times, primarily for comparison purposes, one needs to create several aligned graphs in different subregions of a single figure. Just like in MATLAB, this can be done with the `subplot` command, which will return an array of subplots together with the figure handle. The following script shows how to use this command and also how to set the size or rescale a figure and save it to a file. The resulting image is displayed in figure C.6.

```
x = np.linspace(0,2*np.pi,50)
y1 = np.sin(x)
y2 = np.cos(x)
fig, ax = plt.subplots(1,2) # call subplot function
ax[0].plot(x,y1,'k',linewidth=2) # first subplot in array ax
ax[0].set_title('sin(x)')
ax[1].plot(x,y2,'k',linewidth=2) # second subplot in ax
ax[1].set_title('cos(x)')
fig.suptitle('Subplots',size=10,weight='bold') # sets title
fig = plt.gcf()
fig.set_size_inches(11,5)
fig.savefig("pySubPlots.pdf")
```

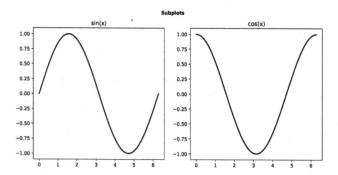

Figure C.6: Subplots created with MatPlotLib.

C.4.7   Newton's method for systems

Finally, here is a simple Python code that implements Newton's method for the system of three equations with three unknowns (5.10). The code also exemplifies a traditional way of documenting Python functions, with a short header that describes the functionality immediately following the declaration line. This Newton code is then called from a different function that sets up the system to be solved and the corresponding Jacobian matrix.

```python
def NewtonSys(F, J, x, eps):
    """
    Solves nonlinear system F(x)=0 by Newton's method.
    J is the Jacobian matrix.
    """
    maxIter = 100
    for k in range(maxIter):
        res = - F(x)
        resNorm = np.linalg.norm(res, ord = 2)
        if resNorm < eps:
            break
        delta = np.linalg.solve( J(x), res)
        x = x + delta
    # Loop ended, return computed values
    return x, k
```

```python
def runNewtonSys():
    """
    Creates the system for Newton's method.
    """
    def F(x):
        return np.array(
            [x[0]**2 + x[1]**2 + x[2]**2 - 1.0,
             x[0]**2 + x[1]**2 - x[2]**2,
             x[0] -x[1] ] )

    def J(x):
        return np.array(
            [[2*x[0], 2*x[1], 2*x[2]],
             [2*x[0], 2*x[1], -2*x[2]],
             [1.0, -1.0, 0.0] ])

    x, n = NewtonSys(F, J, x=np.array([0.1, 0.2, 0.3]), eps=
                     0.0001)
    print(n) # Number of iterations performed
    print(x) # The solution

runNewtonSys() # Runs the example
```

Appendix D

Introduction to Julia

The Julia programming language is a more recent addition to the set of high-level languages which can easily tackle computational problems. Its syntax is quite similar to MATLAB, which makes the transition between the two frameworks relatively easy.

D.1 The Problem of Choice

A relative newcomer on the scene of high-level languages, Julia was initially designed by a group with roots at MIT and Harvard. It takes a different approach from other high-level, dynamically typed languages: instead of being an interpreted language, it uses pre-compilation to potentially speed up actual computations. When using Julia, the needed libraries can be compiled ahead of time; while this compilation requires a large overhead, it only needs to be done once. Codes that use these libraries will run considerably faster than the corresponding codes written in interpreted languages such as MATLAB or even Python. On the other hand, compilation is done under the hood; thus, if a code uses a Julia library that is not pre-compiled, it will be compiled before the code is run. In other aspects, Julia does behave very much like interpreted languages, with their advantage of ease of use. It does have a command-line interface, allows scripts and can easily benefit from interfaces that allow creation of work documents, such as the Jupyter notebook. Julia's user base has recently seen tremendous increase, seeming to indicate that it will become a mainstream language in the near future.

D.2 Julia Basics

The easiest way to use Julia is in an interactive REPL environment, just like for Python and MATLAB. This can either be installed locally or opened in a web browser, and can be made to interact with the Jupyter notebook interface,

DOI: 10.1201/9780429262876-D

the latter offering a way to create documents that contain code that can be run as a script, together with text and equations. There exist a number of free resources that can assist in learning Julia for numerical work, of which the free alternative textbook [16] definitely stands out. Since the coverage there is quite extensive, only a relatively short review is offered below.

D.3   Julia Code Examples

The following script implements Newton's method in Julia as a function:

Julia Script D.1: Newton's Method

```julia
function newtonMethod(f::Function, fprime::Function, x0, alpha,
    N)
    for k = 1:N
        x = x0 - f(x0)/fprime(x0)
        if abs(f(x)) < alpha || abs( x - x0 ) < alpha
            return println("Root: $x; Iteration number: $k")
        end
        x0 = x
    end
    y = f(x)
    println("Newton does not converge: current $x with function
        value $y")
end
```

The arguments `f` and `fprime` to the function `newtonMethod` are explicitly declared to be of type `Function`. This is making the code safer since Julia type checks the arguments. However, when writing Julia code the usual policy is to avoid this explicit declaration. The function `newtonMethod` can then be called in the following way (the output is also shown below as a Julia comment, introduced by the `#` character):

Julia Script D.2: Newton's Method

```julia
# The function could be declared as:
# function newtonMethod(f, fprime, x0, alpha, N)
# It can be called in the following way:
newtonMethod( x -> x^2-5, x->2*x, 5, 10^(-5), 20 )

#Root: 2.2360688956433634; Iteration number: 4
```

Plotting functions in Julia is just as simple as in MATLAB, although there are quite a few plotting packages one can choose from. Simple uses of the `Plots` and the PyPlot packages, which produce figures D.1 and D.2, respectively, are shown below. The scripts also exemplify use of the dot element-wise operations, which have a slightly different but similar syntax with MATLAB. Be aware that when a plotting package is used the first time in a Julia session, pre-compilation usually takes some time. As of the time of this writing this can be frustrating. It is thus best to perform all plotting chores in one Julia session or use a pre-compiled image of the plotting packages.

Julia Script D.3: Using the Plots package

```julia
# Plot a curve using Plots;
# The Plots package uses a number of backends: pyplot, gr,
# unicodeplots and plotlyjs
using Plots
x = range(-2, 2, length = 100 )
y = sin.(x)
plot(x, y)
```

Julia Script D.4: Using the PyPlot package

```julia
using PyPlot
x = range( 0, stop = 2*pi, length=100 )
y = x.*sin.(x);
z = x.*cos.(x);
plot( x, y, color="red", linewidth=2.0, linestyle="--",label="x
    sin(x)")
plot( x, z, color="blue", linewidth=1.0, linestyle="-",label="x
    cos(x)")
legend(loc="lower left")
title("Two functions plot")
```

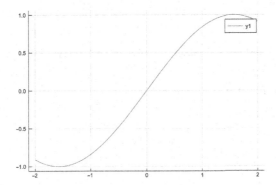

Figure D.1: Plot of the function $f(x) = \sin(x)$ with Julia's Plots package.

Following is a simple Julia code that implements Newton's method for a system and runs it on the particular case in equation 5.10. Notice that the arrays use square brackets to access the elements, similar to Python.

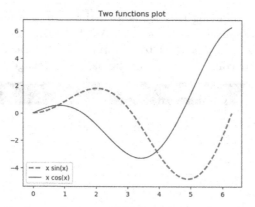

Figure D.2: Multiple function plot with Julia.

```julia
Julia Script D.5: Newton's Method for Systems

using LinearAlgebra

function newtonSys(f, J, x, alpha, N)
    for k = 1:N
        res = - f(x)
        resNorm = norm(res)
        if abs(resNorm) < alpha
            return x
        end
        delta = J(x) \ res
        x = x + delta
    end
    println("Newton's method did not converge yet")
end

f(x) = [  x[1]^2 + x[2]^2 + x[3]^2 - 1.0,
          x[1]^2 + x[2]^2 - x[3]^2,
          x[1] - x[2]   ]
J(x) = [
             2*x[1] 2*x[2] 2*x[3];
             2*x[1] 2*x[2] -2*x[3];
             1.0 -1.0 0.0
           ]
x = [0.1, 0.2, 0.3]
newtonSys(f, J, x, 10^(-5), 100)
```

Finally, here is an implementation of Hermite interpolation that replicates the MATLAB code in 6.5. It exemplifies the way to build a sequence of points in Julia that reproduces the MATLAB command `linspace` with the `range` function and the subsequent conversion of this sequence to an array using `collect`.

Julia Script D.6: Hermite Interpolation in Julia

```julia
function hermite( x, xData, yData, ypData)
    s = 0
    for k = 1:length(xData)
        sq = lagrange(k, x, xData)^2
        xk = xData[k]
        s = s + yData[k] * ( 1 - 2*dL(k, xData)*(x-xk) ) * sq
        s = s + ypData[k] * (x - xk) * sq
    end
    return s
end

function dL( k, xData)
    val = 0
    for j = 1:(k-1)
        val = val + 1.0 / ( xData[k] - xData[j] )
    end
    for j = (k+1):length(xData)
        val = val + 1.0 / ( xData[k] - xData[j] )
    end
    return val
end

function lagrange(k, x, xData)
    p = 1.0
    for j=1:(k-1)
        p = p * (x - xData[j]) / (xData[k] - xData[j])
    end
    for j=(k+1):length(xData)
        p = p * (x - xData[j]) / (xData[k] - xData[j])
    end
    return p
end

xData = collect( range(0, 4*pi, length=5) )
yData = xData .* sin.(xData)
dy = sin.(xData) + xData .* cos.(xData)
xPlot = collect( range(0, 4*pi, length = 100) )
yInt = collect( range(0, 1, length = 100) )
for j = 1:length(xPlot)
    x = xPlot[j]
    yInt[j] = hermite(x, xData, yData, dy)
end

# Plotting
using Plots
gr() #Set the backend to GR
plot(xPlot, yInt)

### This saves the figure in png format (pdf also possilbe)
savefig("hermitePlot.png")
```

Appendix E

Hints and Answers for Selected Exercises

Chapter 2

Ex. 9 The symmetry of the matrix can readily be checked by inspection. To show that is positive-definite, one can compute the eigenvalues. The characteristic equation for the matrix can be found to be

$$\lambda^2 - 5\lambda + 5 = 0,$$

hence the two eigenvalues are:

$$\lambda_1 = \frac{5 - \sqrt{5}}{2}, \quad \lambda_2 = \frac{5 + \sqrt{5}}{2},$$

which are both positive. There are, however, easier ways to determine that the matrix is SPD. One such way is to perform Gauss elimination and check that all the resulting pivots are positive. For larger matrices, it may be even faster to check their positive-definiteness using Sylvester's criterion. In this case, one checks that the determinants of upper-left matrices that are formed starting from the left uppermost entry in A and increasing the rank each time by extending one column to the right and one row below are all positive. In this case there are only two such matrices, the first one being the first entry in A, for which the determinant is 2, the other one being A itself, for which the determinant is 5. Notice that a matrix A cannot be positive-definite if $a_{11} \le 0$.

Chapter 3

Ex. 4 You can start by creating a vector v containing the quantities that need to be added using the colon operator:

```
>> v = @(n) 1 ./ (1:n) ./ (  (1:n) + 2 )
```

Please notice the repeated use of ./, which is due to the fact that multiplication and division have the same precedence. Using a definition as the following would not produce the desired output:

DOI: 10.1201/9780429262876-E

```
>>  v = @(n) 1 ./ (1:n) .* (  (1:n) + 2 )
```

In the next step, one can simply use `cumsum` to obtain the desired function:

```
g = @(n) cumsum( 1 ./ (1:n) ./ ( (1:n) + 2) ) < 3/4
```

Ex. 5 You can use the ambivalence of the `true`/`false` boolean values to define this function as follows:

```
>>  f = @(x) ( x <= 0 ) * ( x+1 ) + ( x > 0 ) * sin(x)
```

Chapter 4

Ex. 9 A recursive function for the task might look as follows:

```
function res = recFib( n )
if n == 1
  res = 0;
elseif n == 2
  res = 1;
else
    res = recFib( n-1 ) + recFib( n-2 );
end
```

Chapter 5

Ex. 3 One can readily obtain an implementation of the Gauss-Seidel method starting from the Jacobi method script 5.2. Nevertheless, consider carefully the variables that are actually needed. Compared to that script, a Gauss-Seidel method code does allow some storage savings.

Ex. 4 $\rho(G_J) \simeq 1.8182$. The Jacobi method does not converge.

Ex. 5 $\rho(G_S) \simeq 0.8668$. The Gauss-Seidel method converges.

Chapter 6

Ex. 17 Remember to shift the indices used in the mathematical notation to suit the range that is offered in MATLAB. For point (b) for example, one can use a MATLAB vector x with index range `1:N`, where $N = n + 1$. It then follows that the elements in this vector should be calculated as `x(K) = - cos( pi * (K-1) / n )`, where `K=1:N`.

Chapter 7

Ex. 7 Substituting $f(x) = 1$, $f(x) = x$ and $f(x) = x^2$ into the numerical differentiation scheme and assuming the derivatives are exact, we obtain $A = C = 1$ and $B = -2$. It is easy to check that for these values of A, B and C, the scheme is also exact for $f(x) = x^3$. It should come as no surprise that one again obtains the central formula.

Chapter 8

Ex. 12 $x = 5$, $y = 4$, $z = 0$ and $P = 18$.

Ex. 14 $\nabla f(x, y) = (4y, 4x)$, $\nabla g(x, y) = (x/128, y/32)$. $4b = \lambda a/128$, $4a = \lambda b/32$. Thus $\lambda = 256$ and $2b = a$. Since $a^2/256 + b^2/64 = 1$, $a = 8\sqrt{2}$, $b = 4\sqrt{2}$ and the maximum area is $4ab = 256$.

Chapter 9

Ex. 8 A template for the code is provided below. The reader is invited to generalize it to suit the requirements of the problem.

MATLAB Script E.1: Partial solution for Ex. 8, Ch. 9

```
function [ val ] = Ex8Gauss( N, K, f )
if N == 1
    w(1) = 2;
    x(1) = 0;
elseif N == 2
    w(1) = 1;
    w(2) = 1;
    x(1) = - 1/sqrt(3);
    x(2) = - x(1);
else
    disp('Error: No such N coded')
end
val = 0;
for j = 1:N
    val = val + w(j) * Ex8Simpson( x(j), f, K );
end
end

function [ g ] = Ex8Simpson( x, f, K )
a = 2 * x^2;
b = 1 + x^2;
h = (b-a)/K;
% terms with a coefficient of 4
s4 = 0;
for j = 2:2:K
    s4 = s4 + feval( f, x, a + (j-1)*h );
end
% terms with a coefficient of 2
s2 = 0;
for j = 3:2:(K-1)
    s2 = s2 + feval( f, x, a + (j-1)*h );
end
% sum up terms above with first and last values in the sum
g = h/3*( 4*s4 + 2 * s2 + feval( f, x, a) + feval( f, x, b) );
end
```

Chapter 10

Ex. 10 A possible implementation of the BEM for this problem is given in
script E.2. The script also plots the numerical solution versus the exact solution. The reader should experiment with different step sizes and compare this with the FEM.

MATLAB Script E.2: Partial solution for Ex. 10, Ch. 10

```
clear all;
close all;
kappa = -20;
t = 0:0.01:3;
y0 = 3/2;
exact = exp( kappa*t ) .* ( y0-cos(0) ) + cos(t);
figure(1)
plot(t, exact)

hold on
n = 15;
dt = 3/n;
A = eye(n) - eye(n)*dt*kappa;
sub = - ones(n-1,1);
A_sub = diag(sub,-1);
A = A + A_sub;
b = zeros(n,1);
t = dt:dt:3;
b(1) = y0 + dt*(-kappa*cos(t(1)) - sin(t(1)));
for i=2:n
   b(i) = dt*(-kappa*cos(t(i)) - sin(t(i)));
end
y = A \ b;
y = y';
yp = [3/2, y] ;
plot(0:dt:3, yp, '-ro')
hold off
xlabel('t')
ylabel('y')
legend('exact solution', 'BEM solution, \Deltat=0.2')
```

Bibliography

[1] W. E. Boyce and R. DiPrima. *Elementary Differential Equations and Boundary Value Problems*. J. Wiley and Sons; 8th edition, 2005.

[2] Stephen Boyd. *Introduction to Applied Linear Algebra*. Cambridge University Press, 2018.

[3] President's Information Technology Advisory Committee. *Computational Science : Ensuring America's Competitiveness*. 2005.

[4] Allen B. Downey. *Think Python. How to think like a computer scientist*. O'Reilly Media; 2nd edition, 2015.

[5] J. D. Faires and R. Burden. *Numerical Methods*. Thomson Brooks/Cole; 3rd edition, 2003.

[6] Michael T. Heath. *Scientific Computing. An Introductory Survey*. McGraw-Hill, 2002.

[7] Mark H. Holmes. *Introduction to Scientific Computing and Data Analysis*. Springer, 2016.

[8] A. Iserles. *A First Course in the Numerical Analysis of Differential Equations*. Cambridge University Press, 1996.

[9] J. F. James. *A student's guide to Fourier transforms*. Cambridge University Press., 1995.

[10] Daniel T. Kaplan. *Introduction to Scientific Computation and Programming*. Thomson Brooks/Cole, 2004.

[11] Jaan Kiusalaas. *Numerical Methods in Engineering with Python 3*. Cambridge University Press, 2013.

[12] David M. Lane. *Introduction to Statistics*. Online Statistics Education, Rice University, 2020.

[13] Serge Lang. *A first course in Calculus*. Springer; 5th edition., 1998.

[14] Bernard W. Lindgren and Donald A. Berry. *Elementary Statistics*. Macmillan, 1981.

[15] Cleve Moler. *Numerical Computing with MATLAB*. Society for Industrial and Applied Mathematics, 2004.

[16] Giray Okten. *First Semester in Numerical Analysis with Julia*. Florida State University, 2019.

[17] Sheldon Ross. *A first course in Probability*. Prentice-Hall; 6th edition., 2002.

[18] James Stewart. *Calculus*. Brooks Cole; 8th edition., 2015.

Index

Printed in the United States
by Baker & Taylor Publisher Services